Leckie×Leckie
Scotland's leading educational publishers

Practice Papers for SQA Exams

National 5

Maths

© 2018 Leckie & Leckie Ltd
001/29012018

10 9 8 7 6 5 4 3 2 1

ISBN 9780008281632

Published by
Leckie & Leckie Ltd
An imprint of HarperCollins*Publishers*
Westerhill Road, Bishopbriggs, Glasgow, G64 2QT
T: 0844 576 8126   F: 0844 576 8131
leckieandleckie@harpercollins.co.uk
www.leckieandleckie.co.uk

Special thanks to
QBS (layout and illustration); Ink Tank (cover design); Roda
Morrison (copy-edit); Paul Sensecall (proofread); Laura Clegg
(proofread)

A CIP Catalogue record for this book is available from the
British Library.

Acknowledgements
Whilst every effort has been made to trace the copyright
holders, in cases where this has been unsuccessful, or if any
have inadvertently been overlooked, the Publishers would
gladly receive any information enabling them to rectify any
error or omission at the first opportunity.

Printed in Italy by Grafica Veneta SpA

Introduction

Layout of the Book

This book contains practice exam papers, which mirror the actual SQA exam as much as possible. The layout, paper colour and question level are all similar to the actual exam that you will sit, so that you are familiar with what the exam paper will look like.

The Answers section is at the back of the book. A fully worked solution is given for each question so that you can see how the right answer has been arrived at. The solutions are accompanied by a commentary which includes further explanations and advice. There is also an indication of how the marks are allocated and, where relevant, what the examiners will be looking for. Reference is made at times to the relevant sections in Leckie & Leckie's National 5 Mathematics Success Guide.

Revision advice is provided in this introductory section of the book, so please read on!

How to use this Book

The Practice Papers can be used in two main ways:

1. You can complete an entire practice paper as preparation for the final exam. If you would like to use the book in this way, you might want to complete each practice paper under exam-style conditions by setting yourself a time for each paper and answering it as best as possible without using any reference materials or notes. Alternatively, you can answer the practice paper questions as a revision exercise, using your notes to produce a model answer. Your teacher may mark these for you.

2. You can use the Topic Index at the front of this book to find all the questions within the book that deal with a specific topic. This allows you to focus specifically on areas that you particularly want to revise or, if you are midway through your course, it lets you practise answering exam-style questions for just those topics that you have studied.

Revision Advice

Work out a revision timetable for each week's work in advance – remember to cover all of your subjects and to leave time for homework and breaks. For example:

Day	6–6.45 p.m.	7–8 p.m.	8.15–9 p.m.	9.15–10 p.m.
Monday	Homework	Homework	English Revision	Chemistry Revision
Tuesday	Maths Revision	Physics revision	Homework	Free
Wednesday	Geography Revision	Modern Studies Revision	English Revision	French Revision
Thursday	Homework	Maths Revision	Chemistry Revision	Free
Friday	Geography Revision	French Revision	Free	Free
Saturday	Free	Free	Free	Free
Sunday	Modern Studies Revision	Maths Revision	Modern Studies Revision	Homework

Make sure that you have at least one evening free a week to relax, socialise and re-charge your batteries. It also gives your brain a chance to process the information that you have been feeding it all week.

Arrange your study time into one hour or 30-minute sessions with a break between sessions (for example, 6–7 p.m., 7.15–7.45 p.m., 8–9 p.m.) Try to start studying as early as possible in the evening when your brain is still alert. Be aware that the longer you put off starting, the harder it will be to start!

Study a different subject in each session, except for the day before an exam.

Do something different during your breaks between study sessions – have a cup of tea, or listen to some music. Don't let your 15 minutes expand into 20 or 25 minutes, though!

Have your class notes and any textbooks available for your revision to hand as well as plenty of blank paper, a pen, etc. You should take note of any topic area that you are having particular difficulty with as and when the difficulty arises. Revisit that question later once having revised that topic area by attempting further questions from the exercises in your textbook.

Revising for a Maths exam is different from revising for some of your other subjects. Revision is only effective if you are trying to solve problems. You may like to make a list of 'Key Questions' with the dates of your various attempts (successful or not!). These should be questions that you have had real difficulty with.

Key Question	1st Attempt		2nd Attempt		3rd Attempt	
Textbook P56 Q3a	18/2/18	✗	21/2/18	✔	28/2/18	✔
Practice Exam A Paper 1 Q5	25/2/18	✗	28/2/18	✗	3/3/18	
2017 SQA Paper, Paper 2 Q4	27/2/18	✗	2/3/18			

The method for using this list is as follows:

1. Any attempt at a question should be dated.

2. A tick or cross should be entered to mark the success or failure of each attempt.

3. A date for your next attempt at that question should be entered:
   (i) for an unsuccessful attempt – three days later
   (ii) for a successful attempt – one week later.

4. After two successful attempts remove that question from the list (you can assume the question has been learnt!).

Using this method for revising for your Maths exam ensures that your revision is focused on the difficulties you have had and that you are actively trying to overcome those difficulties.

Finally, ignore all (or some of) the advice in this section if you are happy with your current way of studying. Everyone revises differently, so find a way that works for you!

Transfer Your Knowledge

As well as using your class notes and textbooks to revise, these practice papers will also be a useful revision tool as they will help you get used to answering exam style questions. As you work through the questions you may find an example that you haven't come across before. Don't worry! There may be several reasons for this. The question may be on a topic that you have not yet covered in class. Check with your teacher to find out if this is the case. Or it could be that the wording or the context of the question is unfamiliar. This often happens with reasoning questions in the Maths exam. Once you have familiarised yourself with the worked solutions, in most cases you will find that the question is using mathematical techniques which you are familiar with. In either case you should revisit that question later to check that you can solve it.

Trigger Words

In the practice papers and in the exam itself, a number of trigger words will be used in the questions. These trigger words should help you identify a process or a technique that is expected in your solution to that part of the question. If you familiarise yourself with these trigger words, it will help you to structure your solutions more effectively.

Trigger Words	Meaning/Explanation
Evaluate	Carry out a calculation to give an answer that is a value.
Hence	You must use the result of the previous part of the question to complete your solution. No marks will be given if you use an alternative method that does not use the previous answer.
Simplify	This means different things in different contexts: (i) Surds: reduce the number under the root sign to the smallest possible by removing square factors. (ii) Fractions: one fraction, cancelled down, is expected. (iii) Algebraic expressions: get rid of brackets and gather all like terms together.
Give your answer to …	This is an instruction for the accuracy of your final answer. These instructions must be followed or you will lose a mark.
Algebraically	The method you use must involve algebra; i.e. you must solve an equation or simplify an algebraic equation. It is usually stated to avoid trial-and-improvement methods or reading answers from your calculator.
Justify your answer	This is a request for you to clearly indicate your reasoning. Will the examiner know how your answer was obtained?
Show all your working	Marks will be allocated for the individual steps in your working. Steps missed out may lose you marks.

In the Exam

Watch your time and pace yourself carefully. You will find some questions harder than others. Try not to get stuck on one question as you may later run out of time. Instead, return to a difficult question later. Remember that if you have spare time towards the end of your exam, use it to check through your solutions. Mistakes are often discovered through this checking process and can be corrected.

Become familiar with the exam instructions. The practice papers in this book have exam instructions at the front of each exam. Also remember that there is a formuae list to consult. You will find this at the front of your exam paper. Even though these formulae are given to you, it is important that you learn them so that they are familiar to you. If you are continuing with Mathematics next session it will be assumed that these formulae are known in next year's exam!

Read the question thoroughly before you begin to answer it – make sure you know exactly what the question is asking you to do. If the question is in sections (for example, 15a, 15b, 15c, etc.) then it is often the case that answers obtained in the earlier sections will be used in the later sections of that question.

When you have completed your solution read it over again. Is your reasoning clear? Will the examiner understand how you arrived at your answer? If in doubt then fill in more details.

If you change your mind, or think that your solution is wrong, don't score it out unless you have another solution to replace it with. Solutions that are not correct can often gain some of the marks available. Do not miss working out. Showing step-by-step working will help you gain marks even if there is a mistake in the working.

Use these resources constructively by reworking questions later that you found difficult or impossible first time round. Remember that success in a Maths exam will only come from actively trying to solve lots of questions, and only consulting notes when you are stuck. Reading notes alone is not a good way to revise for your Maths exam. Always be active, always solve problems.

Good luck!

Topic Index

Topic	Exam A Paper 1	Exam A Paper 2	Exam B Paper 1	Exam B Paper 2	Exam C Paper 1	Exam C Paper 2
Surds	14		12		15	
Indices	14	1	15	6, 9	16	1
Algebraic Expressions: Brackets	6		5		8	
Algebraic Expressions: Factorising			2		5	
Completing the Square	13			10	13	
Algebraic Fractions	4		8		5	
Number Sequences						13
Gradient Formula	5		11		1	
Circles	15		10			5
Volume of Solids Formulae	1			4		9
Functional Notation	7		6		3, 14	
Equation of a Straight Line	5	7	11	8	17	
Linear Equations/ Inequations	8			17	4	8, 14
Simultaneous Equations	11		14		18	
Changing the Subject	15	3		11	6	
Sketching Quadratic Functions	13		16	14	11	
Equations of Quadratic Functions	9		16	14	11	

Topic	Exam A Paper 1	Exam A Paper 2	Exam B Paper 1	Exam B Paper 2	Exam C Paper 1	Exam C Paper 2
Quadratic Equations/Discriminant		13, 15	1		9, 14	3
Pythagoras's Theorem and Converse		12		16		11, 15
Similar Shapes		6		15		15
Trig Graphs	12		7	12		
Trig Equations		5, 14		12, 13		12
Trig Identities			9			6
Area of a Triangle (Trig)		8		13	10	
Sine Rule				3		4
Cosine Rule		11		7	7	
Bearings Problems (Trig)				3		4
Vectors	3		4	5	12	2
Percentages	10	4		1		2
Proportion	10	9				
Fractions	2		3		2	
Quartiles and Averages		2, 10	13	2		10
Standard Deviation		2		2		10
Scattergraphs and Lines of Best Fit	5		11			

In the answers, there are references to the pages of Leckie & Leckie's National 5 Mathematics Success Guide. These will help you learn more about any topics you might find difficult.

N5 Mathematics

Practice Papers for SQA Exams

Duration — 1 hour 15 minutes

Exam A
Paper 1
Non-calculator

Fill in these boxes and read what is printed below.

Full name of centre

Town

Forename(s)

Surname

Total marks — 50

Attempt ALL questions.

You may NOT use a calculator.

To earn full marks you must show your working in your answers.

State the units for your answer where appropriate.

Use **blue** or **black** ink.

Scotland's leading educational publishers

FORMULAE LIST

The roots of $ax^2 + bx + c = 0$ are $x = \dfrac{-b \pm \sqrt{(b^2 - 4ac)}}{2a}$

Sine rule: $\dfrac{a}{\sin A} = \dfrac{b}{\sin B} = \dfrac{c}{\sin C}$

Cosine rule: $a^2 = b^2 + c^2 - 2bc \cos A$ or $\cos A = \dfrac{b^2 + c^2 - a^2}{2bc}$

Area of a triangle: $A = \dfrac{1}{2} ab \sin C$

Volume of a sphere: $V = \dfrac{4}{3} \pi r^3$

Volume of a cone: $V = \dfrac{1}{3} \pi r^2 h$

Volume of a pyramid: $V = \dfrac{1}{3} Ah$

Standard deviation: $s = \sqrt{\dfrac{\sum (x - \bar{x})^2}{n - 1}}$

or $s = \sqrt{\dfrac{\sum x^2 - \dfrac{(\sum x)^2}{n}}{n - 1}}$, where n is the sample size.

1. The diagram shows a toy spinner.
The metal cone at the base of the spinner has a radius
of 2 centimetres and a height of 6 centimetres as shown.
Calculate the volume of the cone.
Take $\pi = 3{\cdot}14$.

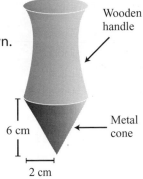

Wooden
handle

6 cm

Metal
cone

2 cm

2

2. Evaluate:

$$2\frac{2}{3} - 1\frac{1}{5} \times \frac{1}{3}$$

3

3. Vectors a and b represent two forces acting on a snooker ball.

$$a = \begin{pmatrix} 3 \\ -2 \\ 3 \end{pmatrix} \text{ and } b = \begin{pmatrix} 5 \\ 2 \\ -1 \end{pmatrix}$$

The resultant force is $a + b$. Calculate the magnitude of this force.
Express your answer as a surd in its simplest form.

3

4. Express as a single fraction in its simplest form:

$$\frac{7}{x} - \frac{3}{x+1}$$

3

5. This scattergraph shows how the students in Iain's Physics class scored in their Physics and Maths exams.

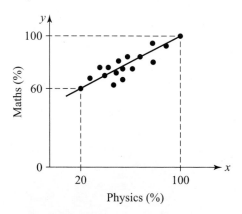

A line of best fit has been drawn.

(a) Find the equation of this line of best fit.

3

(b) Iain missed his Physics exam. If he scored 72 % in his Maths exam, use this equation to estimate a score for his Physics exam.

1

Total Marks **4**

6. Expand and simplify $(3x - 2)(x^2 - x + 2)$.

3

7. Given that $f(x) = \sqrt{10 - x^2}$, find the exact value of $f(-1)$.

2

8. Solve algebraically the inequality:

$$5 < 2 - 3(8 + x)$$

3

9. The diagram below shows part of the graph of $y = (x - k)^2$ where $k > 0$.

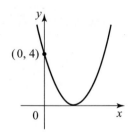

Find the value of k.

2

10. A retailer is offering a 20 % discount on the normal price for all 21-inch 3D televisions.

This is a 21-inch model and is now being sold at a reduced price of £680.
What is the normal price of this television?

3

11. Number Triangles

The rule in a Number Triangle is $A = B + C$.

(a) Use this rule to complete this Number Triangle.

Show that $x + 2y = 3$.

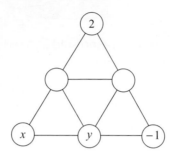

2

(b) The same Number Triangle rule is used for this Number Triangle. Write down another equation with x and y.

2

(c) In both these Number Triangles, x and y have the same values. Find the values of x and y.

3

Total Marks   **7**

12. The graph of $y = p \sin qx°$, $0 \leq x \leq 120$, is shown below.

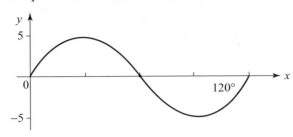

Write down the values of p and q.

2

13. (a) Write $x^2 + 6x + 16$ in the form $(x + p)^2 + q$.

2

(b) Sketch the graph of $y = x^2 + 6x + 16$.

Show clearly:

- The point of intersection with the y-axis

- The coordinates of the turning point.

3

Total Marks 5

14. (a) Evaluate:

$$8^{\frac{2}{3}} - 8^0$$

2

(b) Simplify $\sqrt{2} + \sqrt{8}$.

2

Total Marks **4**

15. The circular clock face on Big Ben is set in a square frame as shown in the diagram.

The tip of the minute hand travels a distance of 27 metres every hour.

Show that the perimeter of the square frame is exactly $\dfrac{108}{\pi}$ metres.

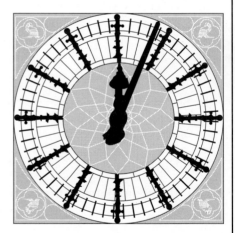

4

[END OF QUESTION PAPER]

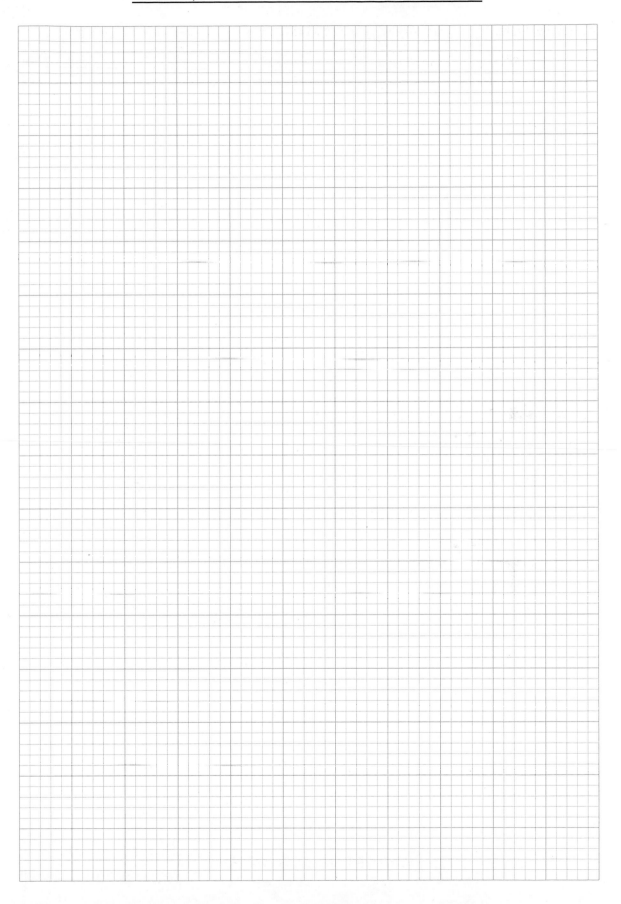

N5 Mathematics

Practice Papers for SQA Exams

Duration — 1 hour 50 minutes

Exam A
Paper 2

Fill in these boxes and read what is printed below.

Full name of centre

Town

Forename(s)

Surname

Total marks — 60

Attempt ALL questions.

You may use a calculator.

To earn full marks you must show your working in your answers.

State the units in your answer where appropriate.

Use **blue** or **black** ink.

Scotland's leading educational publishers

FORMULAE LIST

The roots of $ax^2 + bx + c = 0$ are $x = \dfrac{-b \pm \sqrt{(b^2 - 4ac)}}{2a}$

Sine rule: $\dfrac{a}{\sin A} = \dfrac{b}{\sin B} = \dfrac{c}{\sin C}$

Cosine rule: $a^2 = b^2 + c^2 - 2bc \cos A$ or $\cos A = \dfrac{b^2 + c^2 - a^2}{2bc}$

Area of a triangle: $A = \dfrac{1}{2}ab \sin C$

Volume of a sphere: $V = \dfrac{4}{3}\pi r^3$

Volume of a cone: $V = \dfrac{1}{3}\pi r^2 h$

Volume of a pyramid: $V = \dfrac{1}{3}Ah$

Standard deviation: $s = \sqrt{\dfrac{\sum(x - \bar{x})^2}{n - 1}}$

or $s = \sqrt{\dfrac{\sum x^2 - \dfrac{(\sum x)^2}{n}}{n - 1}}$, where n is the sample size.

1. On 27 August 2003, Mars was 5.6×10^7 km from Earth, its closest approach for 60 000 years.

How long would it take to drive this distance in a car (if this was possible!) at an average speed of 120 kilometres per hour?

Give your answer in days in scientific notation.

4

2. A Quality Control Inspector selects a random sample of seven matchboxes produced by Machine A and records the number of matches in each box:

$$54 \quad 45 \quad 51 \quad 50 \quad 48 \quad 53 \quad 49$$

(a) For the given data calculate:

(i) the mean

1

(ii) the standard deviation

Show clearly all your working.

2

(b) Machine B was also sampled. The data gave a mean of 52 matches and a standard deviation of 1·6 matches. Make two valid comparisons between the results for the two machines.

2

Total Marks **5**

3. Change the subject of the formula

$A = \dfrac{m^2 - n}{2}$ to m.

3

4. In 2001, the Russian Space Station, Mir, was destroyed as it burned up in the upper atmosphere.

It had been losing altitude by 6 % every month.

At the start of December 2000, its altitude was 340 km.

What was its altitude at the start of March 2001?

3

5. Solve the equation:

$3 \sin x° + 2 = 0, 0 \leq x < 360$.

3

6. These two midge-repellent spray cans are mathematically similar.

The smaller can has a height of 8 cm and contains a volume of 48 ml of repellent. The larger can has a height of 10 cm.

Calculate the volume of repellent contained in the larger can.

3

7. (a) A straight line has the equation $2x - 5y + 12 = 0$.

Find the gradient of this line.

2

(b) Find the coordinates of the point where the line crosses the x-axis.

1

Total marks | **3**

8. The instructions for making a hexahexaflexigon start with an 18 cm paper strip of identical equilateral triangles as shown in the diagram below.

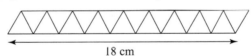

18 cm

Calculate the area of this paper strip.

4

9. The Large Hadron Collider has a circular tunnel with a radius of 4·3 km. In the tunnel, protons are accelerated to almost the speed of light. At this speed they travel 15 km in 51 microseconds (a microsecond is one millionth of a second). In the tunnel there is a detector, Alice, at point A, and a detector, LHC 'B', at point B, and they form an angle of 96° at the centre of the circle as shown.

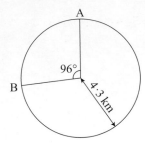

Calculate the time, in microseconds, a proton travelling at this speed takes to travel anticlockwise from Alice to LHC 'B'.

4

10. A biologist was studying the length of time it took bees to return to their hive. These 'out-of-hive' times are shown below in increasing order of time:

$$5 \quad 6 \quad 7 \quad 7 \quad 8 \quad 9 \quad 12 \quad 12 \quad 15$$

$$18 \quad 19 \quad 19 \quad 19 \quad 23 \quad 25 \quad 34 \quad 37 \quad 38$$

(a) Use this data to calculate the semi-interquartile range.

2

(b) For a second hive the semi-interquartile range was found to be 3.5 minutes.

Make a valid comparison between the hives.

1

Total Marks **3**

11. The diagram shows a gear-slider mechanism.

Rod AB = 10·5 cm

Rod BC = 18 cm

angle ABC = 95°

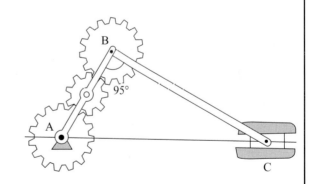

Calculate the length of the slide AC to one decimal place.

4

12. The diagram shows a circular mine shaft with a radius of 5 metres.

A square steel frame is fitted for a lift. Concrete cladding fills the gap between the lift frame and the circular mine shaft.

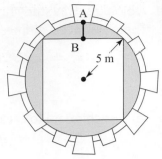

Calculate the greatest width of the concrete cladding (AB in the diagram) giving your answer to 3 significant figures.

5

13. Find the range of values of k such that the equation $2x^2 + x + k = 0$ has real roots.

4

14. The diagram shows a triangular flag with a shaded triangular design.

The flag is in the shape of an isosceles triangle ACD.

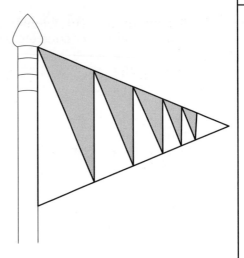

In the design, the largest side of each shaded triangle is at right angles to the lower edge of the flag.

Side AB is 80 cm in length, as shown in the diagram. The tip of the flag forms a 45° angle.

Calculate the length AC of the flag at the flagpole.

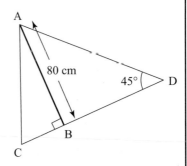

5

15. The diagram shows a swimming pool of width x metres surrounded by a concrete path.

The length of the swimming pool is 2 metres more than its width.

The surrounding concrete path is 1 metre wide along the length and 2 metres wide along the width as shown in the diagram.

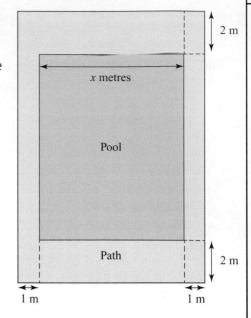

The area of the pool is the same as the area of the concrete path.

(a) Show that $x^2 - 4x - 12 = 0$.

4

(b) Hence find the dimensions of the pool.

3

Total Marks **7**

[END OF QUESTION PAPER]

ADDITIONAL SPACE FOR ANSWERS

Practice Paper B

N5 Mathematics

Practice Papers for SQA Exams

Duration — 1 hour 15 minutes

Exam B
Paper 1
Non-calculator

Fill in these boxes and read what is printed below.

Full name of centre

Town

Forename(s)

Surname

Total marks — 50

Attempt ALL questions.

You may NOT use a calculator.

To earn full marks you must show your working in your answers.

State the units for your answer where appropriate.

Use **blue** or **black** ink.

Leckie×Leckie

Scotland's leading educational publishers

FORMULAE LIST

The roots of $ax^2 + bx + c = 0$ are $x = \dfrac{-b \pm \sqrt{(b^2 - 4ac)}}{2a}$

Sine rule: $\dfrac{a}{\sin A} = \dfrac{b}{\sin B} = \dfrac{c}{\sin C}$

Cosine rule: $a^2 = b^2 + c^2 - 2bc \cos A$ or $\cos A = \dfrac{b^2 + c^2 - a^2}{2bc}$

Area of a triangle: $A = \dfrac{1}{2} ab \sin C$

Volume of a sphere: $V = \dfrac{4}{3} \pi r^3$

Volume of a cone: $V = \dfrac{1}{3} \pi r^2 h$

Volume of a pyramid: $V = \dfrac{1}{3} Ah$

Standard deviation: $s = \sqrt{\dfrac{\sum(x - \bar{x})^2}{n - 1}}$

or $s = \sqrt{\dfrac{\sum x^2 - \dfrac{(\sum x)^2}{n}}{n - 1}}$, where n is the sample size.

MARKS
Do not write in this margin

1. Ailsa attempts to solve the quadratic equation $2x^2 - 3x - 1 = 0$.
    She says the discriminant is negative.
    Do you agree with her? Give a reason for your answer.

    2

2. Factorise:

    $2x^2 + 3x - 2$

    2

3. Evaluate:

    $\frac{2}{3}$ of $\left(1\frac{1}{2} - \frac{1}{3}\right)$

    2

4. $a = \begin{pmatrix} -1 \\ -1 \\ 3 \end{pmatrix}$ and $b = \begin{pmatrix} -2 \\ -3 \\ 7 \end{pmatrix}$ are two vectors.

(a) Write down the components of $3a$.

1

(b) Calculate the components of $3a - 2b$.

1

(c) Calculate the exact value of $|3a - 2b.|$

1

Total Marks 3

MARKS Do not write in this margin

5. Simplify:

$x(2x - 1) - x(2 - 3x)$ **3**

6. $f(x) = \frac{6}{x}$, $x \neq 0$.

(a) Evaluate $f(-3)$. **1**

(b) Given that $f(a) = 3$, find a. **2**

Total Marks 3

7. Part of the graph $y = a \cos bx°$ is shown in the diagram below.

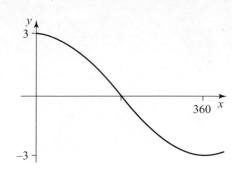

State the values of a and b.

2

8. Express

$$\frac{1}{x - 1} - \frac{3}{x + 2}, \ x \neq 1, x \neq -2$$

as a single fraction in its simplest form.

3

9. Simplify:

$$\frac{\sin^2 x°}{\tan^2 x°}$$

Show your working.

2

10. In the diagram shown below:

- ABC is a tangent to the circle centre O

- The lines AO and BO when extended meet the circle at D and E

- Angle OAB = 40°.

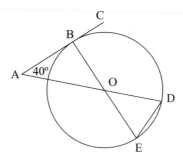

Calculate the size of angle ODE.

3

11. The manager of a car insurance company investigates whether the number of telephone operators he employs is related to the number of complaints his company receives on that day.

The results of his investigation are shown in the scattergraph below, which also shows a line of best fit.

The two data points A(2, 40) and B(8, 22) lie on this line of best fit.

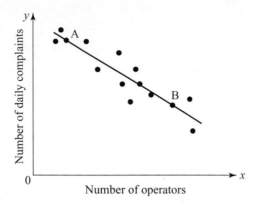

(a) Find the equation of this line of best fit.

3

(b) If he employs six operators, how many complaints should he expect that day?

1

Total Marks 4

12. The area of this rectangle is $10\sqrt{2}\,\text{cm}^2$.

It has breadth $\sqrt{10}\,\text{cm}$.

Calculate the length of the rectangle, expressing your answer as a surd in its simplest form.

$\sqrt{10}$ cm

3

13. The following data gives the ages of the workers on oil rig Alpha:

21	22	24	23	27	23	23
24	21	22	23	23	21	29
23	19	23	21	24	22	23

(a) For this data find:

    (i) The median

1

    (ii) The lower quartile

1

    (iii) The upper quartile

1

(b) For the workers on oil rig Beta, the semi-interquartile range of their ages is 3.5 years. Make an appropriate comment on the distribution of the ages of the workers on the two oil rigs.

2

Total Marks **5**

14. Protons and neutrons are subatomic particles. Inside each atom there is a nucleus made up of protons and neutrons. These subatomic particles are themselves made up of two types of quark: 'up' quarks and 'down' quarks. The electrical charge of a proton or neutron is the sum of the electrical charges of its quarks.

(a) A proton is made of two 'up' quarks and one 'down' quark and has a total electrical charge of 1 unit.

Write down an algebraic equation to illustrate this.

PROTON

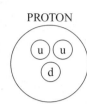

1

(b) A neutron is made of one 'up' quark and two 'down' quarks and has a total electrical charge of zero.

Write down an algebraic equation to illustrate this.

NEUTRON

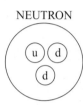

1

(c) Find the electrical charge of an 'up' quark. Show clearly your reasoning.

2

Total Marks 4

15. (a) Evaluate $9^{-\frac{1}{2}}$

2

(b) Simplify $\sqrt{t} \times t^2$

2

Total Marks 4

16.

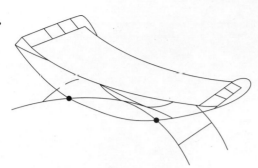

The sketch shows a hammock. Each side of the hammock consists of a metal frame in the shape of two identical parabolas.

Here is a diagram of the frame.

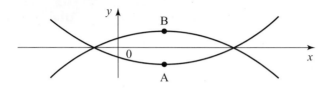

One of these parabolas has equation $y = (x - 2)^2 - 9$. The x-axis is an axis of symmetry for the two parabolas.

(a) State the coordinates of:

    (i)   The turning point A   2

    (ii) The turning point B   1

(b) Find the equation of the parabola whose equation was not given.   2

Total Marks   **5**

[END OF QUESTION PAPER]

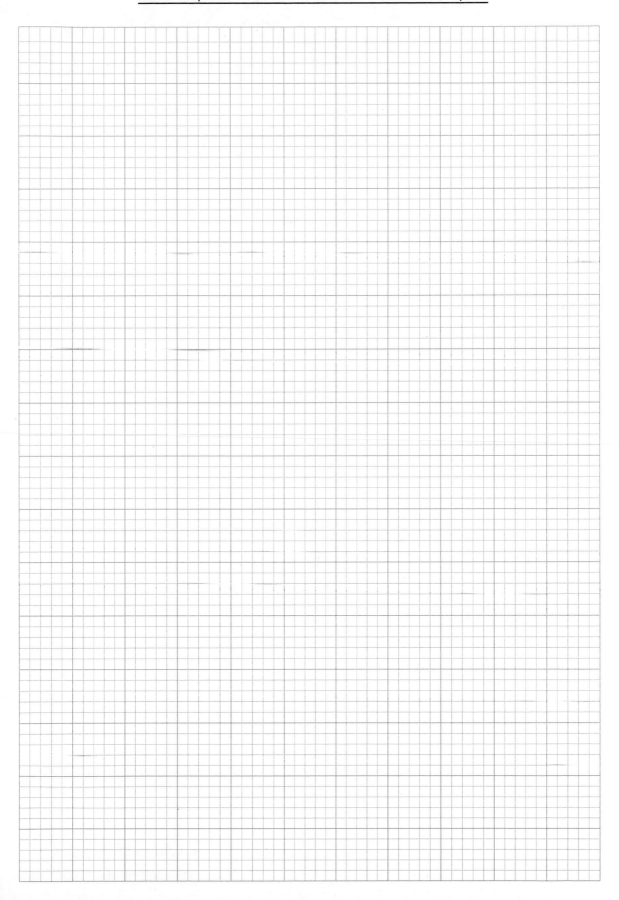

N5 Mathematics

Practice Papers for SQA Exams

Duration — 1 hour 50 minutes

Exam B
Paper 2

Fill in these boxes and read what is printed below.

Full name of centre

Town

Forename(s)

Surname

Total marks — 60

Attempt ALL questions.

You may use a calculator.

To earn full marks you must show your working in your answers.

State the units for your answer where appropriate.

Use **blue** or **black** ink.

FORMULAE LIST

The roots of $ax^2 + bx + c = 0$ are $x = \dfrac{-b \pm \sqrt{(b^2 - 4ac)}}{2a}$

Sine rule: $\dfrac{a}{\sin A} = \dfrac{b}{\sin B} = \dfrac{c}{\sin C}$

Cosine rule: $a^2 = b^2 + c^2 - 2bc\cos A$ or $\cos A = \dfrac{b^2 + c^2 - a^2}{2bc}$

Area of a triangle: $A = \dfrac{1}{2}ab\sin C$

Volume of a sphere: $V = \dfrac{4}{3}\pi r^3$

Volume of a cone: $V = \dfrac{1}{3}\pi r^2 h$

Volume of a pyramid: $V = \dfrac{1}{3}Ah$

Standard deviation: $s = \sqrt{\dfrac{\sum(x - \overline{x})^2}{n - 1}}$

or $s = \sqrt{\dfrac{\sum x^2 - \dfrac{(\sum x)^2}{n}}{n - 1}}$, where n is the sample size.

1. The total emissions of greenhouse gases by the USA in 2007 amounted to the equivalent of 7·2 million tonnes of carbon dioxide. If the annual increase in emissions is 1·2 %, calculate the total amount of emissions of greenhouse gases by the USA expected in 2010.

   Give your answer in millions of tonnes to 2 significant figures.

   4

2. The amounts (in £) spent by a sample of six diners at a restaurant one Saturday evening were:

   £44, £47, £38, £97, £40, £52

   (a) Find the mean amount spent.

   1

   (b) Find the standard deviation of the amounts spent.

   2

   (c) On a weekday evening at the same restaurant, the standard deviation of the amounts spent was £8·40.

   Make one valid comparison between the amounts spent by diners on a weekday evening and a Saturday evening.

   1

   Total Marks 4

3. On this map of Fife, Leven lies due west of Elie.

Cupar is 11·9 km from Leven on a bearing of 004°.

Cupar is on a bearing of 320° from Elie.

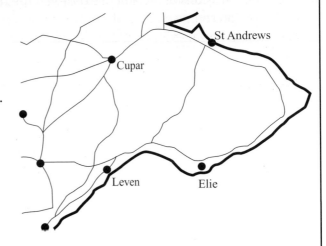

How far is Elie from Leven? Do not use a scale drawing.

4

4. The local 'Thai Cuisine' restaurant uses two types of frying pan: the traditional wok in the shape of a hemisphere and a normal cylindrical pan. The measurements are shown in this diagram:

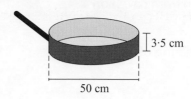

3·5 cm

31 cm

50 cm

Which container has the larger capacity?

5

5. The diagram shows a rectangle ABCD divided into two squares by line PQ.

$\overrightarrow{AD}$ represents vector u and $\overrightarrow{PD}$ represents vector v.

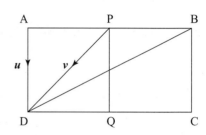

(a) Express $\overrightarrow{PA}$ in terms of u and v.

1

(b) Express $\overrightarrow{BD}$ in terms of u and v.

1

Total Marks **2**

6. The distance from Earth to our nearest star, Proxima Centauri, is $4 \cdot 014 \times 10^{16}$ metres.

Light travels $9 \cdot 461 \times 10^{12}$ kilometres in 1 year.

How many years does it take the light from Proxima Centauri to reach Earth?

2

7. In triangle ABC, AB $= 10$ km, AC $=12$ km and angle BAC $= 10°$.

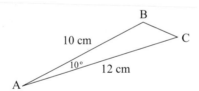

Calculate the length of BC.

3

8. (a) The diagram below shows a part of a straight line graph with equation $3d + t = 15$.

Find the gradient of this line.

2

(b) State the coordinates of point A.

1

Total marks 3

9. Express $\dfrac{1}{(\sqrt[3]{a})^2}$ in the form a^n.

2

10. Express $\dfrac{2-a}{a^2} + \dfrac{1}{a}$, $a \neq 0$

as a single fraction in its simplest form.

3

11. Change the subject of the formula

$P = \sqrt{9 - Q^2}$ to Q.

3

12. Part of the graph of $y = 5\sin x° - 1$ is shown below.

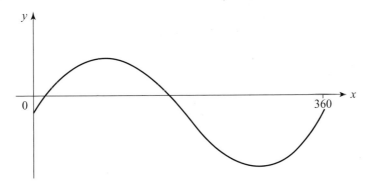

Calculate the x-coordinates of the two points where the graph cuts the x-axis.

4

13. The diagram shows the design for a triangular pendant.

The area of the central triangle PQR is 4·2 cm².

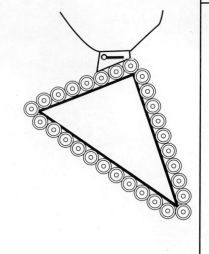

PR = 2·5 cm

PQ = 3·4 cm

Calculate the size of the acute angle QPR at the top of the pendant.

3

14. The diagram below shows the graph of a quadratic function with the equation

$$y = k(x - p)(x - q) \text{ where } p < q$$

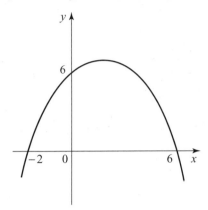

The graph cuts the x-axis at the points $(-2, 0)$ and $(6, 0)$ and cuts the y-axis at the point $(0, 6)$.

(a) Write down the values of p and q. **2**

(b) Calculate the value of k. **2**

(c) Find the coordinates of the maximum turning point of the function. **2**

Total Marks **6**

15. These two organ pipes are mathematically similar in shape.

The larger pipe is 240 cm in length and the smaller pipe is 180 cm in length.

The volume of the larger pipe is 43 litres.

Calculate the volume of the smaller pipe. Give your answer correct to the nearest litre.

3

16. This spanner head is circular with a symmetrical 5-sided gap.

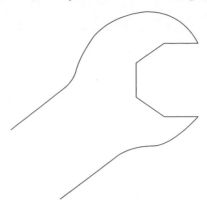

The diagram below shows the dimensions of the circular head.
C is the centre of the circle which has diameter 5 cm.
The head measures 4·8 cm across as shown.

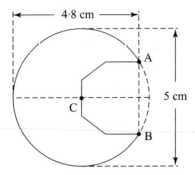

Calculate the width of the gap AB.

4

17. Marcus travelled from St Andrews to Thurso in two stages.

(a) In the first stage of his journey he covered 180 miles in x hours.

Find, in terms of x, his average speed.

1

(b) He covered the 60 miles of the second stage of his journey in 2 hours less time than the first stage.

Find an expression for his average speed for the second stage of his journey.

1

(c) His average speed on both stages of his journey was the same. Calculate the time taken for the whole of his journey.

3

Total Marks   **5**

[END OF QUESTION PAPER]

Practice Paper C

N5 Mathematics

Practice Papers for SQA Exams

Duration — 1 hour 15 minutes

Exam C
Paper 1
Non-calculator

Fill in these boxes and read what is printed below.

Full name of centre

Town

Forename(s)

Surname

Total marks — 50

Attempt ALL questions.

You may NOT use a calculator.

To earn full marks you must show your working in your answers.

State the units for your answer where appropriate.

Use **blue** or **black** ink.

Scotland's leading educational publishers

FORMULAE LIST

The roots of $ax^2 + bx + c = 0$ are $x = \dfrac{-b \pm \sqrt{(b^2 - 4ac)}}{2a}$

Sine rule: $\dfrac{a}{\sin A} = \dfrac{b}{\sin B} = \dfrac{c}{\sin C}$

Cosine rule: $a^2 = b^2 + c^2 - 2bc \cos A$ or $\cos A = \dfrac{b^2 + c^2 - a^2}{2bc}$

Area of a triangle: $A = \dfrac{1}{2} ab \sin C$

Volume of a sphere: $V = \dfrac{4}{3} \pi r^3$

Volume of a cone: $V = \dfrac{1}{3} \pi r^2 h$

Volume of a pyramid: $V = \dfrac{1}{3} Ah$

Standard deviation: $s = \sqrt{\dfrac{\sum(x - \bar{x})^2}{n - 1}}$

or $s = \sqrt{\dfrac{\sum x^2 - \dfrac{(\sum x)^2}{n}}{n - 1}}$, where n is the sample size.

1. Show that the gradient of AB where A(k, k^2) and B(1, k) is k.

3

2. Evaluate:

$$\frac{2}{3} \div 1\frac{1}{3}$$

2

3. Given that $f(x) = x(2 - x)$, evaluate $f(-1)$.

2

4. Solve the inequality $3 + 2x < 4(x + 1)$.

3

5. (a) Factorise $9y^2 - 4$.

1

(b) Hence simplify:

$$\frac{9y^2 - 4}{15y - 10}$$

2

Total Marks 3

6. $P = \dfrac{W}{4A}$

Change the subject of this formula to A.

2

7. The triangle below shows the distances between Cupar (C), Leven (L) and St Andrews (S).

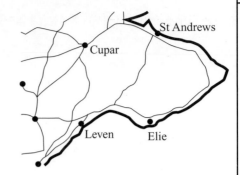

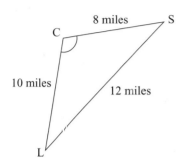

$CS = 8$ miles

$CL = 10$ miles

$LS = 12$ miles

Show that $\cos C = \dfrac{1}{8}$.

3

MARKS
Do not
write in the
margin

8. Expand and simplify $(2 + x)(3 - 2x - x^2)$.

3

9. Determine the nature of the roots of the function $f(x) = 3x^2 - x - 5$.

2

10. In triangle ABC:

- BC $= 2$ cm

- AC $= 5$ cm

- $\sin C = \dfrac{7}{10}$

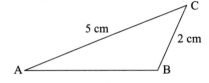

Calculate the area of the triangle.

2

11. The graph below shows a parabola with an equation of the form $y = (x + a)^2 + b$. The line $x = 2$ is the axis of symmetry of the parabola.

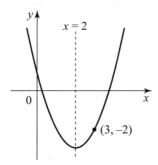

(a) State the value of a.

1

The point $(3, -2)$ lies on the parabola.

(b) Calculate the value of b.

2

Total marks | **3**

12.

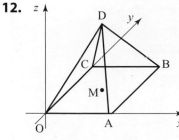

OABCD is a square-based pyramid. Sides OA and OC are placed on the x-axis and y-axis, respectively, as shown.

O is the origin.

D lies vertically above M, the centre of the square base.

A has coordinates $(8, 0, 0)$.

(a) Find the coordinates of M.

1

(b) If DM = OA find the coordinates of D.

1

(c) $\overrightarrow{AD}$ represents vector w where $w = \begin{pmatrix} -4 \\ 4 \\ 8 \end{pmatrix}$

Calculate $|w|$, the magnitude of w.

Express your answer as a surd in its simplest form.

2

Total Marks 4

13. Write $x^2 - 5x + \dfrac{1}{4}$ in the form $(x - p)^2 + q$.

2

14. Part of the graphs $y = f(x)$ and $y = g(x)$ are shown where:

$f(x) = 4 + 3x - x^2$

$g(x) = 10 - 2x$

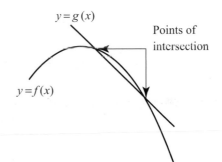

Find the x-coordinates of the two points of intersection by solving the equation $f(x) = g(x)$.

3

15. Simplify fully:

$$\sqrt{\left(\sqrt{18}\right)^2 + \left(\sqrt{6}\right)^2}$$

2

16. Express in its simplest form:

$$x^3 \times (x^{-1})^{-2}$$

2

17. A baby measured 20 inches in length at birth.
The baby's length was recorded at regular intervals and a graph plotted.

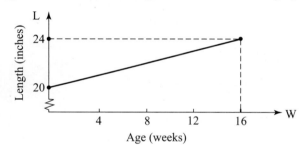

The graph is linear and shows that at 16 weeks the baby's length is 24 inches.

Find the equation of the straight line graph in terms of W and L.

3

18. Carrie is a tea blender for a large international tea company.

She has recently bought quantities of Kenyan and Rwandan tea and has created two different blends from these teas. They are made in 5 kg packets.

Blend A: 2 kg of Kenyan and 3 kg of Rwandan costing 6·10 euros.

Blend B: 3 kg of Kenyan and 2 kg of Rwandan costing 5·90 euros.

Let the cost of Kenyan be x euros per kg and the cost of Rwandan be y euros per kg.

(a) Write down an algebraic equation to illustrate the make up and cost of Blend A.

1

(b) Write down a similar equation for Blend B.

1

(c) She creates a third blend as follows:

Blend C: 4 kg of Kenyan and 1 kg of Rwandan.

Find the cost of 5 kg of Blend C.

4

Total Marks **6**

[END OF QUESTION PAPER]

ADDITIONAL SPACE FOR ANSWERS

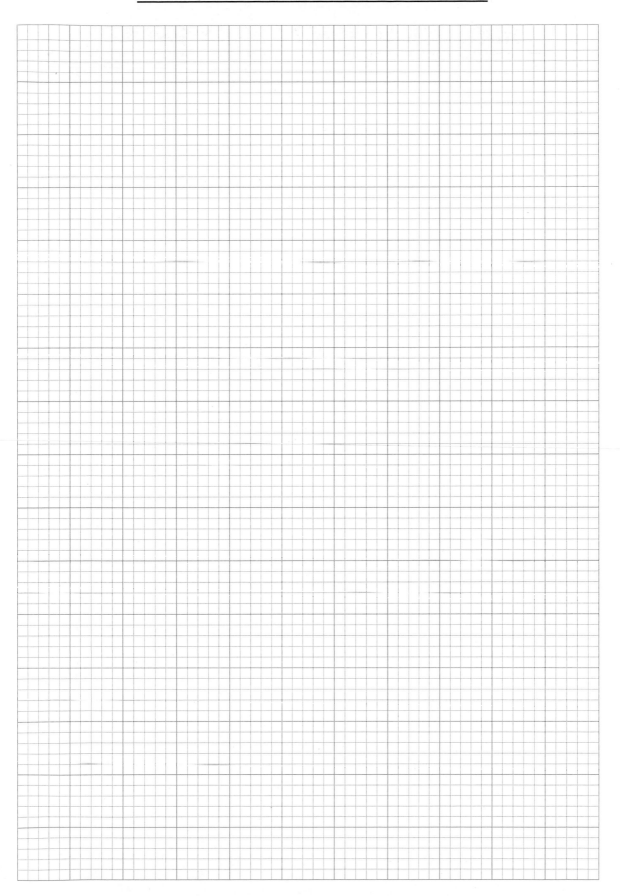

N5 Mathematics

Practice Papers for SQA Exams

Duration — 1 hour 50 minutes

Exam C
Paper 2

Fill in these boxes and read what is printed below.

Full name of centre

Town

Forename(s)

Surname

Total marks — 60

Attempt ALL questions.

You may use a calculator.

To earn full marks you must show your working in your answers.

State the units for your answer where appropriate.

Use **blue** or **black** ink.

Scotland's leading educational publishers

FORMULAE LIST

The roots of $ax^2 + bx + c = 0$ are $x = \dfrac{-b \pm \sqrt{(b^2 - 4ac)}}{2a}$

Sine rule: $\dfrac{a}{\sin A} = \dfrac{b}{\sin B} = \dfrac{c}{\sin C}$

Cosine rule: $a^2 = b^2 + c^2 - 2bc \cos A$ or $\cos A = \dfrac{b^2 + c^2 - a^2}{2bc}$

Area of a triangle: $A = \dfrac{1}{2} ab \sin C$

Volume of a sphere: $V = \dfrac{4}{3} \pi r^3$

Volume of a cone: $V = \dfrac{1}{3} \pi r^2 h$

Volume of a pyramid: $V = \dfrac{1}{3} Ah$

Standard deviation: $s = \sqrt{\dfrac{\sum(x - \bar{x})^2}{n-1}}$

or $s = \sqrt{\dfrac{\sum x^2 - \dfrac{(\sum x)^2}{n}}{n-1}}$, where n is the sample size.

1. In 2009 scientists at Stanford University broke the record for the smallest writing. They wrote 'SU'. Each letter had width 3×10^{-8} cm.

   In letters this size the entire Bible could be written in a line 1.05×10^{-1} cm long.

   Calculate the number of letters in the Bible. Give your answer in scientific notation.

   2

2. Mr Middleton was informed by 'Petcare' that his annual premium for his pet insurance had been increased by $6\frac{1}{2}$ %.

   His annual premium is now £87·82.

   Calculate his annual premium before this increase.

   3

3. Solve the equation:

   $3x^2 - x - 5 = 0$

   Give your answer correct to one decimal place.

   4

4. From Dublin the bearing of Edinburgh is 035° and the bearing of London is 118°.

From Edinburgh the bearing of London is 158°.

The distance from Dublin to Edinburgh is 343 km.

Calculate the distance from Edinburgh to London.

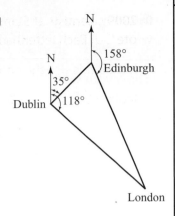

5

5. A cylindrical plastic water pipe has a uniform thickness of $2\frac{1}{2}$ cm as is shown in the cross-sectional diagram below.

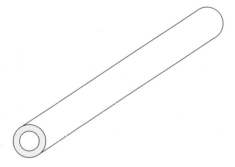

The outside diameter of the pipe is 30 cm.

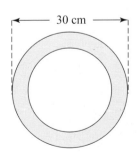

Calculate the area of plastic in the cross-section.

3

6. Simplify

$\sin x° \cos x° \tan x° + \cos^2 x°$

Show your working.

3

7. Among the shapes found in a wooden puzzle game are the following:

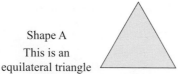

Shape A
This is an
equilateral triangle

Shape B
This is an
isosceles triangle

These two shapes fit perfectly together to form a larger triangle as shown below:

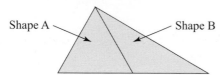

Shape A

Shape B

Is this larger triangle right-angled?

Show all your working and justify your answer

3

8. (a) Solve the equation

$$\frac{2}{3} - \frac{3x}{4} = \frac{x}{2}$$

Give your answer as a fraction in its simplest form.

3

(b) The two lines with equations $y = \frac{2}{3} - \frac{3x}{4}$ and $y = \frac{x}{2}$ intersect at the point P.

Find the value of the y-coordinate of the point P.

1

Total Marks **4**

9. The diagram on the right shows a plumb-line used by bricklayers to ensure that their constructions are vertical.

The weight at the bottom is in the shape of a cone and is made of metal.

Here are the dimensions of the cone:

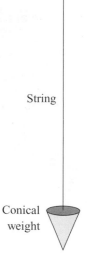

String

Conical weight

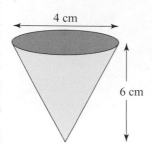

4 cm

6 cm

(a) Calculate the volume of this conical weight.

Give your answer correct to three significant figures.

3

(b) The weight is redesigned into the shape of half of a cylinder as shown in the diagram on the right. The same metal is used for this new weight and has the same volume as the old conical weight.

Calculate the height of this new weight.

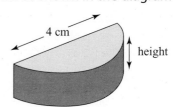

4 cm

height

3

Total Marks   **6**

10. A ream of paper contains 500 sheets.

A sample of five reams was checked and the number of sheets recorded as:

503   504   497   495   506

For this data the mean number of sheets is 501.

(a) Calculate the standard deviation.

Show clearly all your working.

3

(b) It was felt by the producer that there was too much variation in the number of sheets in the sample.

The machine that bundled the reams was adjusted and a new sample was tested. The new mean and standard deviation for this sample were 502 sheets and 3·5 sheets, respectively. Did the adjustment produce less variation?

Give a reason for your answer.

1

Total Marks **4**

11. An aircraft fuselage has a circular cross-section of diameter 4·6 metres.

The passenger compartment floor is 1·2 metres above the lowest point of the cargo compartment as shown below:

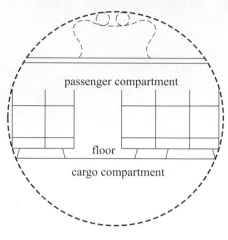

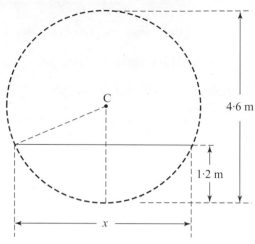

In the diagram above C is the centre of the cross-section.

(a) Calculate x, the width of the passenger compartment floor.

4

(b) The passenger compartment ceiling is the same width as the floor.
How high above the floor is it?

1

Total Marks **5**

12. The diagram shows part of the graph $y = \cos x°$.

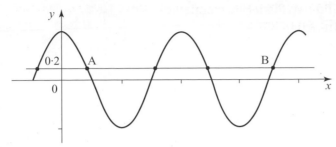

The line $y = 0.2$ cuts the graph shown at five points.

Find the x-coordinates of A and B, the 2nd and 5th of these five points.

3

13. Here is a number pattern:

$2 + 1 = 1 \times 3 - 2 \times 0$

$4 + 1 = 2 \times 4 - 3 \times 1$

$6 + 1 = 3 \times 5 - 4 \times 2$

(a) Write down the line of this pattern.

1

(b) Write down the nth line of this pattern.

2

(c) Hence show algebraically that this pattern is always true.

1

Total Marks 4

14. A company is hiring an Excavator machine.

They are considering two Plant Hire companies,
'Earthmove' and 'Trenchers'.

The hire rates for these companies are as follows:

'Earthmove'	**'Trenchers'**
Delivery charge: £64	Delivery charge: £28
Hourly rate: £30	Hourly rate: £34

(a) Calculate the cost of a 2-hour hire from each of the companies.

1

(b) For a 20-hour hire which company is cheaper?

1

(c) For each company find a formula for the cost of a '*n*' hour hire.

2

(d) 'Trenchers' claim that they are the cheaper of the two companies.

Find algebraically the greatest number of hours of hire for this claim to be true.

2

Total Marks 6

15. This diagram shows the supporting wooden structure for a roof.

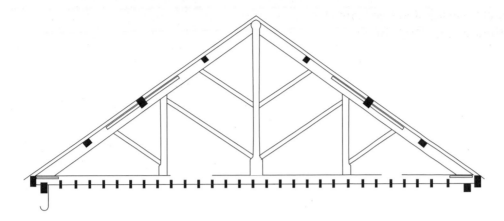

The diagram on the right shows the dimensions of one of the central sections bounded by the two vertical posts AG and DF.

The 'web truss' BC is 2·7 m in length and is parallel to the other 'web truss' DE.

Section AC of the 'King post' AG is 3 m in length and the lower section CE is 2 m in length.

EH is a horizontal line parallel to the 'bottom chord' GF with DH 2·5 m in length.

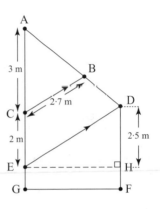

Calculate the distance, GF, between the two vertical posts.

5

[END OF QUESTION PAPER]

ADDITIONAL SPACE FOR ANSWERS

Answers

Answers to Exam A

Exam A: Paper 1

Q1.

$\text{Volume} = \dfrac{1}{3}\pi r^2 h$

with $\pi = 3 \cdot 14$, $r = 2$ and $h = 6$

so $\text{Volume} = \dfrac{1}{3} \times 3 \cdot 14 \times 2^2 \times 6$ ✓

$= \dfrac{1}{3} \times 6 \times 2^2 \times 3 \cdot 14$

$= 2 \times 4 \times 3 \cdot 14$

$= 8 \times 3 \cdot 14$

$= 25 \cdot 12 \text{ cm}^3$ ✓

2 marks

Substitution

- The formula $V = \dfrac{1}{3}\pi r^2 h$ is given to you on the formulae sheet during your exam
- The first mark is given for the correct substitution of the numbers into the formula
- You do not have access to a calculator in Exam 1 so all calculations must be done on your answer paper. This is why the value $\pi = 3 \cdot 14$ is given for this calculation. You must not use any other value if you are asked to use this one.

Calculation

- The second mark is for correctly calculating the volume
- No marks are awarded for rounding in this question. This is generally the case in a question that does not specify the accuracy required for the answer.

SUCCESS GUIDE: page 41

Q2. $2\dfrac{2}{3} - 1\dfrac{1}{5} \times \dfrac{1}{3}$

$= \dfrac{8}{3} - \left(\dfrac{6}{5} \times \dfrac{1}{3}\right)$ ✓

$= \dfrac{8}{3} - \dfrac{6}{15}$ ✓

$= \dfrac{40}{15} - \dfrac{6}{15}$

$= \dfrac{40 - 6}{15}$

$= \dfrac{34}{15}$

$= 2\dfrac{4}{15}$ ✓

3 marks

Order of operations

- Multiplication is done before subtraction

Multiplication of fractions

- Change mixed fractions to 'top-heavy' fractions $1\dfrac{1}{5} = \dfrac{6}{5}$
- Multiply 'top' numbers; multiply 'bottom' numbers $\dfrac{6}{5} \times \dfrac{1}{3} = \dfrac{6 \times 1}{5 \times 3} = \dfrac{6}{15}$

SUCCESS GUIDE: page 122

Subtraction of fractions

- A common denominator is needed. 3 and 15 both divide exactly into 15.
- $\dfrac{8}{3} = \dfrac{8 \times 5}{3 \times 5} = \dfrac{40}{15}$ multiply 'top' and 'bottom' by 5
- Change back to mixed fractions.

SUCCESS GUIDE: page 123

Q3.

$$a+b = \begin{pmatrix} 3 \\ -2 \\ 3 \end{pmatrix} + \begin{pmatrix} 5 \\ 2 \\ -1 \end{pmatrix} = \begin{pmatrix} 8 \\ 0 \\ 2 \end{pmatrix} \quad \checkmark$$

$$|a+b| = \sqrt{8^2 + 0^2 + 2^2} = \sqrt{64+4} = \sqrt{68} \quad \checkmark$$

$$\sqrt{68} = \sqrt{4 \times 17} = \sqrt{4} \times \sqrt{17} = 2\sqrt{17} \quad \checkmark$$

3 marks

Addition
- Correct components will gain 1st mark

Magnitude
- You have to know that the magnitude of a vector with components $\begin{pmatrix} p \\ q \\ r \end{pmatrix}$ is given by $\sqrt{p^2 + q^2 + r^2}$

SUCCESS GUIDE: page 116

Simplification
- Remember to look for factors that are square numbers: 4, 9, 16, …
- $\sqrt{17}$ cannot be simplified further since 17 is a prime number and has no factors other than 1 and 17.

SUCCESS GUIDE: page 9

Q4. $\dfrac{7}{x} - \dfrac{3}{x+1}$

$$= \frac{7(x+1)}{x(x+1)} - \frac{3x}{x(x+1)} \quad \checkmark$$

$$= \frac{7(x+1) - 3x}{x(x+1)}$$

$$= \frac{7x+7-3x}{x(x+1)}$$

$$= \frac{4x+7}{x(x+1)} \quad \checkmark$$

2 marks

Subtraction of algebraic fractions
- A common denominator is needed: multiply x and $x+1$ together
- Always multiply the 'top' and the 'bottom' by the same expression: $\dfrac{7 \times (x+1)}{x \times (x+1)}$ and $\dfrac{3 \quad \times x}{x+1 \times x}$

Simplify
- The numerators are subtracted and then simplified.

SUCCESS GUIDE: page 34

Q5. (a) Two points on the line are
A(20, 60) and B(100, 100)

so $m_{AB} = \dfrac{100-60}{100-20} = \dfrac{40}{80} = \dfrac{1}{2}$ $\quad \checkmark$

now use $y - b = m(x - a)$

with $m = \dfrac{1}{2}$ and (a, b) as (20, 60)

so the equation is $\quad \checkmark$

$$y - 60 = \frac{1}{2}(x - 20)$$

$$\Rightarrow y - 60 = \frac{1}{2}x - 10$$

$$\Rightarrow y = \frac{1}{2}x + 50 \quad \checkmark$$

3 marks

Gradient
- Use the formula $m = \dfrac{y_2 - y_1}{x_2 - x_1}$

SUCCESS GUIDE: pages 36, 37

Substitution
- Alternatively $y = mx + c$ may be used giving $60 = \dfrac{1}{2} \times 20 + c$ (leading to $c = 50$)

Equation
- Alternative forms are: $2y = x + 100$ or $2y - x = 100$.

SUCCESS GUIDE: pages 56–58

Q5. (b) The Maths scores are the y values
so let $y = 72$
Now solve $72 = \frac{1}{2}x + 50$

$\Rightarrow \frac{1}{2}x = 22 \Rightarrow x = 44$ ✓
His estimate is 44 %

1 mark

Solve equation
- Be careful to read the descriptions on the axes. A common mistake here is to substitute $x = 72$.
SUCCESS GUIDE: pages 57, 58

Q6. $(3x - 2)(x^2 - x + 2)$
$= 3x(x^2 - x + 2) - 2(x^2 - x + 2)$ ✓
$= 3x^3 - 3x^2 + 6x - 2x^2 + 2x - 4$ ✓
$= 3x^3 - 5x^2 + 8x - 4$ ✓

3 marks

Start the expansion
- There are other strategies e.g. $(3x - 2)x^2 - (3x - 2)x + (3x - 2) \times 2$

Complete the expansion
- Remove all of the brackets

Simplification
- Gather the like terms and simplify.
SUCCESS GUIDE: page 24

Q7. $f(x) = \sqrt{10 - x^2}$
so $f(-1) = \sqrt{10 - (-1)^2}$ ✓
$= \sqrt{10 - 1}$
$= \sqrt{9} = 3$ ✓

2 marks

Substitution
- Replace x by -1

Evaluation
- Note that squaring a negative gives a positive result.
SUCCESS GUIDE: page 47

Q8. $5 < 2 - 3(8 + x)$
$\Rightarrow 5 < 2 - 24 - 3x$ ✓
$\Rightarrow 5 < -22 - 3x$
$\Rightarrow 3x < -22 - 5$ ✓
$\Rightarrow 3x < -27$
$\Rightarrow x < -9$ ✓

3 marks

Expand brackets
- Be careful with the sign of the terms: -3×8 and $-3 \times (+x)$ are both negative

Collect like terms
- Add $3x$ and substract 5 from both sides of the equation

Solve for x
- Divide both sides by 3.
SUCCESS GUIDE: page 50

Q9. For $(0, 4)$ use $x = 0$ and $y = 4$.
Substitute these values in
$y = (x - k)^2$
$\Rightarrow 4 = (0 - k)^2$ ✓
$\Rightarrow 4 = (-k)^2$
$\Rightarrow 4 = k^2$
$\Rightarrow k = 2(k > 0)$ ✓

2 marks

Substitution
- Point $(0,4)$ lies on the graph so $x = 0$ and $y = 4$ satisfy the equation of the graph.

Solve for k
- $k = -2$ ruled out as a possibility since the question states $k > 0$.
SUCCESS GUIDE: page 71

Q10. Normal price (100%) has been reduced by 20% to 80%:

80% $\longleftrightarrow$ £680   ✓

1% $\longleftrightarrow$ $\dfrac{680}{80}$ ✓

100% $\longleftrightarrow$ $\dfrac{680}{80} \times 100$ ✓

$= £850$ ✓

3 marks

Percentages
- The final price is given *after* a reduction so a 'proportion' method is required.
- Knowing 80%, divide by 80 to find 1% and multiply by 100 to find 100%.

SUCCESS GUIDE: pages 119, 120

Q11. (a) This gives: ✓

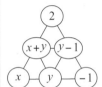

$x + y + y - 1 = 2$

$x + 2y - 1 = 2$

$x + 2y = 2 + 1$

$x + 2y = 3$ ✓

2 marks

Understanding the pattern
- The example A = B + C tells you that the top circle is the sum of the two below it:
  x + y and y + -1

Set up an equation
- Following the pattern for the top triangle leads to an equation:
  $(x + y)$ + $(y - 1)$ = 2

Q11. (b) This gives: ✓

$y + 2 + 2 + 2x = 13$

$y + 4 + 2x = 13$

$x + 2x = 13 - 4$

$y + 2x = 9$ ✓

2 marks

Set up an equation
- Following the pattern for the top triangle leads to an equation:
  $(x + y)$ + $(y - 1)$ = 2

Simplify the equation
- Letters on the left, numbers on the right.

Q11. (c) Solve simultaneous equations

$\left.\begin{array}{l} x + 2y = 3 \\ 2x + y = 9 \end{array}\right\}$ $\begin{array}{l} \times 2 \longrightarrow 2x + 4y = 6 \\ \longrightarrow 2x + y \;= 9 \end{array}$ ✓

$\underset{\text{subtract}}{} \;\; 3y = -3$

so $y = -1$ ✓

Put $y = -1$ into $2x + y = 9$

so $2x - 1 = 9 \Rightarrow 2x = 10$

$\Rightarrow x = 5$ ✓

3 marks

Simultaneous equations
- You will recognise this from having two equations and two unknowns (x and y)
- Line up the equations x's under x's and y's under y's with numbers on the right.

Calculation of values
- Aim for one equation with one unknown
- Substitute the first calculated value back into either of the two equations
- Check using the other equation. Use $y = -1$ and $x = 5$ in $x + 2y$ to get $5 + 2x(-1) = 3$.

SUCCESS GUIDE: page 61

Q12. The amplitude is 5 so
$$p = 5 \quad ✓$$

From 0° to 360° there would be
3 cycles so $q = 3$ ✓

2 marks

Amplitude
- $y = \sin x°$ has a max or min value of 1 or −1 so $y = 5 \sin x°$ has a max or min value of 5 or −5, true for this graph.

Period
- $y = \sin x°$ has period 360° − 1 cycle every 360°. So $y = \sin 3x°$ has period 120° − 3 cycles every 360°; true for this graph.

SUCCESS GUIDE: page 94

Q13. (a) Compare $x^2 + 6x + 16$
with $(x + 3)^2 = x^2 + 6x + 9$ ✓
so $x^2 + 6x + 16$
$$= (x + 3)^2 - 9 + 16 \quad ✓$$
$$= (x + 3)^2 + 7$$

2 marks

Value of p
- The coefficient of x is halved: $\frac{1}{2}$ of $6 = 3$

Value of q
- When $(x + 3)^2$ is multiplied out, the constant term 9 is subtracted and the required constant 16 added.

SUCCESS GUIDE: page 28

Q13. (b)

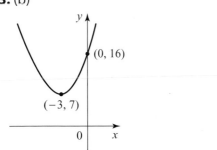

(0, 16)

(−3, 7)

✓
✓
✓

3 marks

Parabolic shape
- $y = x^2$ is shifted left and moved up

y-intercept
- set $x = 0$ in $y = x^2 + 6x + 16$ to get $y = 16$

Turning point
- The shift to $y = x^2$ is 3 left and 7 up.

SUCCESS GUIDE: page 71

Q14. (a) $8^{\frac{2}{3}} - 8^0$

$$= (^3\sqrt{8})^2 - 1 \quad ✓$$
$$= 2^2 - 1 = 4 - 1 = 3 \quad ✓$$

2 marks

Zero index
- $a^0 = 1$ Any non-zero number raised to the power zero gives 1.

Fractional indices
- $a^{\frac{m \leftarrow \text{power}}{n \leftarrow \text{root}}}$ so $a^{\frac{m}{n}} = (^n\sqrt{a})^m$ and in this case: $8^{\frac{2 \leftarrow \text{squared (power 2)}}{3 \leftarrow \text{cube root}}}$

SUCCESS GUIDE: page 18

Q14. (b)
$$\sqrt{2} + \sqrt{8}$$
$$= \sqrt{2} + \sqrt{4 \times 2}$$
$$= \sqrt{2} + \sqrt{4} \times \sqrt{2} \quad ✓$$
$$= \sqrt{2} + 2\sqrt{2} = 3\sqrt{2} \quad ✓$$

Simplifying surds
- Always attempt to get the smallest number under the square root by factorising out squares: 4,9,16,...

Adding surds
- Compare $= \sqrt{2} + 2\sqrt{2} = 3\sqrt{2}$ with $x + 2x = 3x$.

SUCCESS GUIDE: pages 9, 11

2 marks

Q15. In 1 hour the tip of the minute hand travels the whole circumference of the clock face ✓

so $\pi D = 27$ ✓

$\Rightarrow D = \dfrac{27}{\pi}$ ✓

The diameter, D, of the clock face is the same length as the side of the square frame

so perimeter $= 4 \times D$

$= 4 \times \dfrac{27}{\pi}$

$= \dfrac{108}{\pi}$ metres ✓

4 marks

Strategy
- Evidence that you knew to use the Circumference formula to set up an equation will gain you the strategy mark.

Set up equation
- $C = 2\pi r$ or $C = \pi D$: the second version is the best one for this question.
SUCCESS GUIDE: page 38

Change the subject
- Divide both sides by π

Calculation
- 4 sides make up the perimeter, so $\times 4$
- The working $4 \times \dfrac{27}{\pi}$ is essential especially since the answer $\dfrac{108}{\pi}$ is given.

Exam A: Paper 2

Q1. Use $T = \dfrac{D}{S}$

so $T = \dfrac{5 \cdot 6 \times 10^{7}}{120}$ ✓

$= 466666 \cdot 66...$ hours ✓

$= \dfrac{466666 \cdot 66...}{24}$ days

$= 19444 \cdot 44...$

$= 1 \cdot 9 \times 10^{4}$ days ✓
(to 2 significant figures)

4 marks

D, S, T formula

- You are asked to find the Time taken. Covering T in the triangle gives $\dfrac{D}{S}$, the Distance divided by the Speed.

Calculator
- $5 \cdot 6 \times 10^{7}$ is entered into your calculator like this:

$\boxed{5}\;\boxed{\cdot}\;\boxed{6}\;\boxed{\times 10^{x}}\;\boxed{7}\;\boxed{=}$

Conversion
- Change hours to days: divide by 24.

Scientific notation
- Convert your answer to scientific notation:

$$a \times 10^{n}$$

a number between 1 and 10, in this case 1·9

The decimal point in 1·9 moves 4 places to the right so $n = 4$

SUCCESS GUIDE: pages 18, 19

Q2. (a)(i)

$$\text{Mean} = \frac{54+45+51+50+48+53+49}{7}$$

$$= \frac{350}{7} = 50 \qquad \checkmark$$

1 mark

Mean

- $\text{Mean} = \dfrac{\text{sum of numbers}}{\text{number of numbers}} = \dfrac{\sum x}{n} = \bar{x}$

Q2. (a)(ii)

x	$x - \bar{x}$	$(x - \bar{x})^2$
54	4	16
45	−5	25
51	1	1
50	0	0
48	−2	4
53	3	9
49	−1	1
	$\sum(x - \bar{x})^2 = 56$	

$\checkmark$

$\checkmark$

$$s = \sqrt{\frac{\sum(x - \bar{x})^2}{n-1}} = \sqrt{\frac{56}{6}} \doteq 3\cdot1$$

2 marks

Squared deviations

- This mark is gained for the correct last column in the table: the squared values of the deviations from the mean.

Substitution & calculation

- The formula $s = \sqrt{\dfrac{\sum(x - \bar{x})^2}{n-1}}$ is given to you in the exam on your formulae page.
- This mark is for correctly substituting $\sum(x-x)^2 = 56$ and $n - 1 = 6$ into the formula and completing the calculation.
- Using the other formula gives $\sum x = 350$, $\sum x^2 = 17556$ and $n = 7$, resulting in $s \doteq 3\cdot1$.
- If you know how to use the STAT mode on your calculator to check $s = 3\cdot1$ then you should do this. Remember just writing the answer down from the calculator with no working will not gain you the marks.

SUCCESS GUIDE: pages 130, 131

Q2. (b)

On average there are more matches in Machine B's boxes (mean = 52 is greater than mean = 50 for machine A). $\qquad \checkmark$

There is much more variation in the number of matches in Machine A's boxes than those from Machine B ($s = 3\cdot1$ for Machine A is greater than $s = 1\cdot6$ for Machine B). $\qquad \checkmark$

2 marks

1ˢᵗ statement

- This concerns the 'average' contents. Back up your statement with the statistics (52 and 50).

2ⁿᵈ statement

- This concerns the distribution about the mean. A larger standard deviation means more variation about the mean.

Q3. $A = \dfrac{m^2 - n}{2}$

$(\times 2)\ (\times 2)$ ✓

$\Rightarrow 2A = m^2 - n$

$(+n)\quad (+n)$ ✓

$\Rightarrow 2A + n = m^2$

So $m^2 = 2A + n$

$\Rightarrow m = \sqrt{2A + n}$ ✓

3 marks

1st step
- Get rid of the fraction: multiply both sides of the equation by 2.

2nd step
- Try to isolate the expression involving the subject m: add n to both sides.

Last step
- The 'inverse' of squaring is 'square rooting' so take the square root of both sides.
SUCCESS GUIDE: pages 64, 65

Q4. Reduction factor is 0·94 and is applied three times. ✓

So final altitude

$= 340 \times 0\cdot94 \times 0\cdot94 \times 0\cdot94$ ✓

$= 282\cdot39... \doteq 282$ km

(to nearest km) ✓

3 marks

Reduction factor
- Reducing by 6% is equivalent to calculating 94%, i.e. multiplying by 0·94.
SUCCESS GUIDE: page 119

Time calculation
- Shorter version is $340 \times 0\cdot94^3$

Rounding
- Unless accuracy is stated in the question any correct rounding is acceptable.

Q5. $3 \sin x^\circ + 2 = 0$

$\Rightarrow 3 \sin x^\circ = -2 \Rightarrow \sin x^\circ = -\dfrac{2}{3}$ ✓

x° is in the 3rd or 4th quadrant; the 1st quadrant angle is 41·8°

So $x = 180 + 41\cdot8$

$= 221\cdot8$ ✓

or

$x = 360 - 41\cdot8$

$= 318\cdot2$ ✓

3 marks

Rearrangement
- The 1st step is to find the value of $\sin x^\circ$

1st value
- $\sin x^\circ$ is negative: use

$$\begin{array}{c|c} S & A \\ \hline T & C \end{array}\ x \quad \text{or}$$

to determine the quadrants (3rd and 4th)
- 1st quadrant angle is obtained from $\sin^{-1}\left(\dfrac{2}{3}\right)$. Do not use the negative with $\boxed{\sin^{-1}}$
- 3rd quadrant angle is always $180° +$ (1st quadrant angle)

2nd value
- quadrant angle is always $360° -$ (1st quadrant angle)
- Use this diagram to remember how to get the different quadrant angles (q is the 1st quadrant angle).

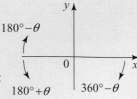

$180°-\theta$

$180°+\theta \qquad 360°-\theta$

SUCCESS GUIDE: pages 96, 97

Q6. Scale factor for lengths
$$= \tfrac{10}{8} = \tfrac{5}{4}(= 1 \cdot 75) \qquad ✓$$

Scale factor for volumes
$$= (\tfrac{5}{4})^3 (= 1 \cdot 75^3) \qquad ✓$$

Volume of larger can
$$= 50 \times (\tfrac{5}{4})^3 = 93 \cdot 75 \text{ ml} \qquad ✓$$

3 marks

Scale factor for lengths
- You are to find the **larger** volume so the **larger** height goes on the top of the fraction (scale factor >1)

Scale factor for volumes
- This is the cube of the length scale factor

Calculation
- $93\tfrac{3}{4}$ ml is an alternative answer.

SUCCESS GUIDE: pages 87–88

Q7. (a) $2x - 5y + 12 = 0$
$$\Rightarrow -5y = -2x - 12$$
$$\Rightarrow y = \tfrac{-2}{-5}x - \tfrac{12}{(-5)}$$
$$\Rightarrow y = \tfrac{2}{5}x + \tfrac{12}{5} \qquad ✓$$
So the gradient $= \tfrac{2}{5}$ ✓

2 marks

Rearrangement
- You are rearranging the equation into the form $y = mx + c$. An alternative method is $2x + 12 = 5y$ and then divide both sides by 5.

Gradient
- For the arrangement $y = mx + c$ the gradient is m, the coefficient of x.

SUCCESS GUIDE: page 52

Q7. (b) Set $y = 0$ for the y-axis intercept

so $2x - 5y + 12 = 0$

gives $2x + 12 = 0$
$$\Rightarrow 2x = -12$$
$$\Rightarrow x = -6$$
The required point is $(-6, 0)$ ✓

1 mark

Coordinates
- The value $y = 0$ is substituted into the equation of the line.
- The coordinates are required so $x = -6$ is not enough to gain this mark.

Q8. There are 9 triangles arranged along one 18 cm side so each triangle has side length 2 cm.

Area of one triangle ✓
$$= \tfrac{1}{2} \times 2 \times 2 \times \sin 60° \qquad ✓$$
$$= 2 \sin 60°$$

Area of the whole strip
$$= 18 \times 2 \sin 60°$$
$$= 3 \sin 60° = 31.2 \text{ cm}^2 \qquad ✓$$

4 marks

Angles and lengths
- These triangles have three angles of 60° and side length of 2 cm.

Area of one triangle
- Here you are using the formula
$$A = \tfrac{1}{2}\,ab \sin C$$
with $a = b = 2$ and $C = 60°$

SUCCESS GUIDE: page 104

Area of strip
- The strip consists of 18 triangles
- Any correct approximation will be accepted e.g. 31 cm² or 31·18 cm²

Q9. Length of arc AB

$$= \frac{96}{360} \times \text{circumference} \quad ✓$$

$$= \frac{96}{360} \times \pi \times 8 \cdot 6 = 7 \cdot 204...\text{km} \quad ✓$$

So 15 km $\longleftrightarrow$ 51 microseconds

$$\Rightarrow 1 \text{ km} \longleftrightarrow \frac{51}{15} \text{ microseconds}$$

$$\Rightarrow 7 \cdot 204... \longleftrightarrow \frac{51}{15} \times 7 \cdot 204 \text{ microseconds}$$

$$= 24 \cdot 496 \quad ✓$$

The proton takes 24·5 microseconds to travel
from A to B in the tunnel. ✓

4 marks

Strategy
- The angle at the centre of the circle is in direct proportion to the length of the arc. So we have
- 96° $\longleftrightarrow$ arc AB
- 360° $\longleftrightarrow$ whole circumference
- This leads to $\frac{96}{360}$ as the required fraction of the circumference that gives the length of arc AB.
- This mark is for the appearance of $\frac{96}{360}$ or equivalent.

Arc length
- $C = \pi D$ where $D = 2 \times$ radius
  $= 2 \times 4\cdot3$ km

Method
- A proportion problem: Divide by 15 then multiply by 7·204…. Evidence that you knew this will gain you this mark.

Solution
- Correct calculation gains the final mark.
SUCCESS GUIDE: page 38

Q10. (a) There are 18 measurements. Divide them, in increasing order, into two groups of nine:
(5............15)(18............38)
Divide the nine lower measurements, in increasing order, into two groups:

(5 6 7 7)   8   (9 12 12 15)

The lower quartile (Q_1) = 8
Divide the nine upper measurements, in increasing order, into two groups:
(18 19 19 19)   23   (25 34 37 38)

The upper quartile (Q_3) = 23 ✓
semi-interquartile range

$$= \frac{Q_3 - Q_1}{2} = \frac{23 - 8}{2} = 7.5 \quad ✓$$

2 marks

Lower quartile and Upper quartile
- Another way of thinking about the lower quartile is that it is the median of the lower half of the measurements.
- The upper quartile can be thought of as the median of the upper half of the measurements.
- Remember that the symbol Q_2 is reserved for the median of the whole set of data.

Semi-interquartile range
- The formula SIQR $= \frac{Q_3 - Q_1}{2}$ is not given to you in the exam so you need to remember it.
SUCCESS GUIDE: pages 127–129

Q10. (b)
The times for the 2nd hive are less varied in their distribution about the median time (semi-interquartile range is 3.5) than for the 1st hive (semi-interquartile range is 7.5 which is greater than 3.5). ✓

1 mark

Statement
- In your statement you must compare the two semi-interquartile ranges 3.5 and 7.5 otherwise you will not gain this mark.
SUCCESS GUIDE: page 128

Q11. Use the cosine rule:
$$b^2 = a^2 + c^2 - 2ac\,\cos B \qquad \checkmark$$

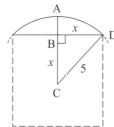

$$= 18^2 + 10 \cdot 5^2 - 2 \times 18 \times 10 \cdot 5 \times \cos 95° \qquad \checkmark$$
$$= 467 \cdot 19... \qquad \checkmark$$
so $b = \sqrt{467 \cdot 19}... = 21 \cdot 61...$
so AC $\doteqdot 21 \cdot 6$ cm (to 1 decimal place) $\qquad \checkmark$

Strategy
- Two sides and the included angle are given so the cosine rule is used.

Substitution
- Correct substitution of the values into the formula will gain you a mark.

Calculation
- In most calculators just key in the complete calculation as written in the solution.

Rounding
- 1 mark is allocated for correct rounding.
SUCCESS GUIDE: page 108

4 marks

Q12.

Triangle BCD (see diagram) is isosceles with BC = BD = x $\qquad \checkmark$ $\checkmark$

By Pythagoras's Theorem: $\qquad \checkmark$
$$x^2 + x^2 = 5^2$$
$$\Rightarrow 2x^2 = 25$$
$$\Rightarrow x^2 = \frac{25}{2} = 12 \cdot 5 \qquad \checkmark$$
So $x = \sqrt{12 \cdot 5}$. AC is a radius
so AC $= 5$ $\qquad \checkmark$
$$AB = AC - BC = 5 - \sqrt{12 \cdot 5}$$
$$= 5 - 3 \cdot 53... = 1 \cdot 464...$$
So AB $\doteqdot 1 \cdot 46$ cm (to 3 significant figures)

Strategy
- Constructing the right-angled triangle and using Pythagoras's Theorem is the crucial strategy.

Pythagoras's Theorem
- Naming BC and BD with letter x helps.

Strategy
- Realisation that AC is a radius, length 5 m, is the 2nd crucial strategy step.

Calculation
- Never round answers until the final answer.

Rounding
- Rounding to 3 significant figures is essential.
SUCCESS GUIDE: pages 76, 84, 85

5 marks

Q13. Compare $2x^2 + x + k = 0$
    with $ax^2 + bx + c = 0$
This gives $a = 2$, $b = 1$ and $c = k$ $\qquad \checkmark$
so Discriminant $= b^2 - 4ac$
$$= 1^2 - 4 \times 2 \times k \qquad \checkmark$$
$$= 1 - 8k$$
For real roots, Discriminant ≥ 0
so $1 - 8k \geq 0$ $\qquad \checkmark$
$$\Rightarrow 1 \geq 8k \Rightarrow \frac{1}{8} \geq k$$
So the range is $k \leq \frac{1}{8}$ $\qquad \checkmark$

Values
- This is for obtaining the values of the coefficients a, b and c.

Discriminant
- This mark is for knowing to use the discriminant $b^2 - 4ac$.

Inequation
- No real roots if discriminant < 0, real roots in all other cases.

Solution
- Add $8k$ to both sides of the inequation then divide both sides by 8.
- Notice that $a \leq b$ is the same as $b \geq a$.
SUCCESS GUIDE: page 74

4 marks

Q14. The large triangle is isosceles

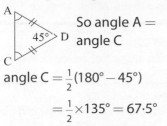

So angle A = angle C ✓

angle C $= \frac{1}{2}(180° - 45°)$

$= \frac{1}{2} \times 135° = 67 \cdot 5°$ ✓

Now use 'SOHCAHTOA' in triangle ABC ✓

$\sin 67 \cdot 5° = \dfrac{80}{AC}$ ✓

$\Rightarrow AC \times \sin 67 \cdot 5° = 80$

$\Rightarrow AC = \dfrac{80}{\sin 67 \cdot 5°} = 86 \cdot 59\ldots$

$\doteq \underline{\underline{86 \cdot 6 \text{ cm}}}$ ✓

5 marks

Strategy
- Evidence that you knew to use 'SOHCAHTOA' in triangle ABC by finding angle C would gain you the strategy mark

Angle sum
- Angle A and angle C between them make up 135° of the 180° angle sum in the large triangle.

Correct ratio
- Working from angle C: the opposite AB is known and the hypotenuse AC is required, so use 'SOH', i.e. sin.

Correct equation
- $\sin 67 \cdot 5 = \dfrac{80}{AC}$ or its equivalent will gain you a further mark.

Calculation
- Multiply both sides of the equation by AC and then divide both sides by sin 67·5°.
SUCCESS GUIDE: pages 89, 90

Q15. (a) Use Area = length × breadth

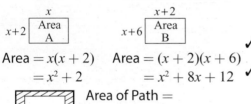

Pool: Outside of Path:

Area $= x(x + 2)$ Area $= (x + 2)(x + 6)$ ✓
$= x^2 + 2$ $= x^2 + 8x + 12$ ✓

Area of Path =
Area B − Area A
$= x^2 + 8x + 12 - (x^2 + 2x)$
$= x^2 + 8x + 12 - x^2 - 2x$
$= 6x + 12$ ✓
Area of Pool = Area of Path
$\Rightarrow x^2 + 2x = 6x + 12$
$\Rightarrow x^2 + 2x - 6x - 12 = 0$
$\Rightarrow x^2 - 4x - 12 = 0$ ✓

4 marks

Starting
- You will gain a mark for either of the two area expressions appearing.

Strategy
- Any correct method for finding an expression for the area of path and then setting up an equation will guarantee the strategy mark.

Simplification
- Simplifying expressions as you proceed through your solution makes the subsequent working easier.

Equation rearrangement
- Since the answer is given ($x^2 - 4x - 12 = 0$) it is important that your working is clear.

Q15. (b) $x^2 - 4x - 12 = 0$
$\Rightarrow (x - 6)(x + 2) = 0$ ✓
$\Rightarrow x = 6$ or $x = -2$
$x = -2$ is not sensible for a length
So $x = 6$ is the only solution. ✓
Pool dimensions:
$6 \text{ m} \times 8 \text{ m}$ ✓

3 marks

Solving a Quadratic
- Each factor could be zero:
$x - 6 = 0$ or $x + 2 = 0$ giving the two solutions $x = 6$ and $x = -2$.

Invalid Solution
- You must state clearly that $x = -2$ makes no sense. There is a mark for this.

Answer
- The dimensions of the pool are x metres by $(x + 2)$ metres with $x = 6$.
SUCCESS GUIDE: pages 67, 68

Answers to Exam B

Exam B: Paper 1

Q1. Compare $2x^2 - 3x - 1 = 0$
with $ax^2 + bx + c = 0$
This gives $a = 2, b = -3$ and $c = -1$
Discriminant $= b^2 - 4ac$
$= (-3)^2 - 4 \times 2 \times (-1)$ ✓
$= 9 + 8 = 17$
This is positive so she is wrong. ✓

2 marks

Substitution
- For correct substitution into $b^2 - 4ac$.

Calculation and statement
- Remember negative $\times$ negative gives a positive result:
$(-3)^2 = -3 \times (-3) = 9$ and $-4 \times 2 \times (-1) = 8$
- You will lose this mark if you fail to mention that 17 is positive (or non-negative).

Q2. $2x^2 + 3x - 2$
$= (2x - 1)(x + 2)$ ✓ ✓

2 marks

Factorisation
- You should always check your answer by multiplying out using 'FOIL'.
SUCCESS GUIDE: pages 26, 27

Q3. $\dfrac{2}{3}$ of $\left(1\dfrac{1}{2} - \dfrac{1}{3}\right)$

$= \dfrac{2}{3}$ of $\left(\dfrac{3}{2} - \dfrac{1}{3}\right)$

$= \dfrac{2}{3}$ of $\left(\dfrac{9}{6} - \dfrac{2}{6}\right) = \dfrac{2}{3} \times \dfrac{9-2}{6}$ ✓

$= \dfrac{2}{3} \times \dfrac{7}{6} = \dfrac{14}{18} = \dfrac{7}{9}$ ✓

2 marks

Subtraction of fractions
- The lowest common denominator is 6 as this is the smallest number 2 and 3 divide into exactly.

Multiplication of fractions
- "Multiply across": $\dfrac{a}{b} \times \dfrac{c}{d} = \dfrac{a \times c}{b \times d}$
- Cancel if possible: in this case divide top and bottom of the fraction by 2 to get $\dfrac{7}{9}$.
SUCCESS GUIDE: pages 122, 123

Q4. (a)

$3a = 3\begin{pmatrix} -1 \\ -1 \\ 3 \end{pmatrix} = \begin{pmatrix} 3\times(-1) \\ 3\times(-1) \\ 3\times 3 \end{pmatrix} = \begin{pmatrix} -3 \\ -3 \\ 9 \end{pmatrix}$ ✓

1 mark

Components
- Each component is trebled.

Q4. (b)

$3a - 2b = \begin{pmatrix} -3 \\ -3 \\ 9 \end{pmatrix} - 2\begin{pmatrix} -2 \\ -3 \\ 7 \end{pmatrix} = \begin{pmatrix} -3+4 \\ -3+6 \\ 9-14 \end{pmatrix} = \begin{pmatrix} 1 \\ 3 \\ -5 \end{pmatrix}$ ✓

1 mark

Components
- Take care with negative signs

$-2\begin{pmatrix} -2 \\ -3 \\ 7 \end{pmatrix}$ $\begin{matrix} \leftarrow\text{positive} \\ \leftarrow\text{positive} \\ \leftarrow\text{negative} \end{matrix}$ $\begin{matrix} -2\times(-2) \\ -2\times(-3) \\ -2\times 7 \end{matrix}$

Q4. (c)

$$|3\boldsymbol{a} - 2\boldsymbol{b}| = \left\| \begin{pmatrix} 1 \\ 3 \\ -5 \end{pmatrix} \right\| = \sqrt{1^2 + 3^2 + (-5)^2}$$

$$= \sqrt{1 + 9 + 25} = \sqrt{35} \quad ✓$$

1 mark

Magnitude
- $(-5)^2 = -5 \times (-5)$ is positive
- $\sqrt{35}$ cannot be simplified as 35 has no square factors

SUCCESS GUIDE: pages 116, 117

Q5. $x(2x - 1) - x(2 - 3x)$

$$= 2x^2 - x - 2x + 3x^2 \quad ✓✓$$

$$= 5x^2 - 3x \quad ✓$$

3 marks

Brackets
- Take care with a negative multiplier:
  $-x(2 - 3x) = -2x + 3x^2$ (2nd term positive).

Simplify
- Gather 'like terms': $2x^2 + 3x^2$
  and $-x - 2x$.

SUCCESS GUIDE: pages 20–22

Q6. (a)

$$f(x) = \frac{6}{x}$$

$$\text{so } f(-3) = \frac{6}{-3} = -2 \quad ✓$$

1 mark

Substitution
- From $f(x)$ to $f(-3)$ involves replacing x by the value -3 and calculating the result.

SUCCESS GUIDE: page 47

Q6. (b)

$$f(a) = 3 \Rightarrow \frac{6}{a} = 3 \quad ✓$$

$$\text{so } 6 = 3a \Rightarrow a = 2 \quad ✓$$

2 marks

Equation
- $f(a)$ is replaced by $\frac{6}{a}$ for the 1st mark.

Solving
- Divide both sides by 3.

Q7. The amplitude of the graph is 3
so $a = 3$. ✓

From $0°$ to $360°$ there is half a cycle so $b = \frac{1}{2}$. ✓

2 marks

Value of a
- a gives the amplitude of the graph.

Value of b
- b gives the number of cycles from $0°$ to $360°$ which in this case is $\frac{1}{2}$. Note the period is $720°$ and so an alternative calculation is $b = \frac{360}{720}$.

SUCCESS GUIDE: pages 94–95

Q8.

$$\frac{1}{x-1} - \frac{3}{x+2}$$

$$= \frac{1 \times (x+2)}{(x-1) \times (x+2)} - \frac{3 \times (x-1)}{(x-1) \times (x+2)} \quad ✓$$

$$= \frac{x+2 - 3(x-1)}{(x-1)(x+2)} \quad ✓$$

$$= \frac{x+2 - 3x+3}{(x-1)(x+2)}$$

$$= \frac{-2x+5}{(x-1)(x+2)} \quad ✓$$

3 marks

Correct denominator
- The common denominator is $(x-1)(x+2)$. The first fraction is multiplied by $\frac{x+2}{x+2}$ and the second one by $\frac{x-1}{x-1}$.

Correct numerator
- $x + 2 - 3(x-1)$ unsimplified earns this mark

Answer
- Simplify the numerator. Sometimes further cancelling occurs but not in this case.
- The numerator could also be written as $5 - 2x$.
SUCCESS GUIDE: page 33

Q9.

$$\frac{\sin^2 x^\circ}{\tan^2 x^\circ}$$

$$= \frac{\sin^2 x^\circ}{\frac{\sin^2 x^\circ}{\cos^2 x^\circ}} \quad ✓$$

$$= \frac{\sin^2 x^\circ \times \cos^2 x^\circ}{\frac{\sin^2 x^\circ}{\cos^2 x^\circ} \times \cos^2 x^\circ}$$

$$= \frac{\sin^2 x^\circ \times \cos^2 x^\circ}{\sin^2 x^\circ}$$

$$= \cos^2 x^\circ \quad ✓$$

2 marks

Correct trig identity
- Use $\tan x^\circ = \frac{\sin x^\circ}{\cos x^\circ}$
- Note that:
$\tan^2 x^\circ = \tan x^\circ \times \tan x^\circ$
$$= \frac{\sin x^\circ}{\cos x^\circ} \times \frac{\sin x^\circ}{\cos x^\circ} = \frac{\sin^2 x^\circ}{\cos^2 x^\circ}$$

Simplification
- Alternatively: dividing by $\frac{\sin^2 x^\circ}{\cos^2 x^\circ}$ is equivalent to multiplying by $\frac{\cos^2 x^\circ}{\sin^2 x^\circ}$. You may have learnt this method.
This leads to: $\sin^2 x^\circ \times \frac{\cos^2 x^\circ}{\sin^2 x^\circ} = \cos^2 x$.
SUCCESS GUIDE: page 99

Q10. The line ABC is a tangent to the circle so angle OBA = 90°
⇒ angle AOB = 50° ✓
(the angles in △OAB add up to 180°)
Angle DOE = 50° (since it is vertically opposite angle AOB) ✓
△DOE is isosceles
(OD = OE are equal radii)
⇒ angle DOE = angle EOD
$$= \tfrac{1}{2}(180° - 50°)$$
$$= \tfrac{1}{2} \times 130° = 65°$$ ✓
Here is a diagram:

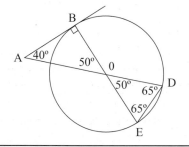

3 marks

Calculation of 1st angle
- You should know that the tangent (line ABC) and the radius (line OB) to the point of contact (point B) are perpendicular.

Calculation of 2nd angle
- These are sometimes called X-shaped angles. They are on opposite sides of the vertex O and are equal.

Final answer
- It is worth checking your answer at this stage using the angles of △DOE : 50° + 65° + 65° = 180°
- Make sure your working and reasoning are clearly shown throughout.

Q11. (a) A(2, 40) B (8, 22)

so $m_{AB} = \dfrac{22-40}{8-2} = \dfrac{-18}{6} = -3$ ✓

now use $y - b = m(x - a)$
with $m = -3$ and $(a, b) = (2, 40)$

The equation is
$y - 40 = -3(x - 2)$ ✓
$\Rightarrow y - 40 = -3x + 6$
$\Rightarrow y = -3x + 46$ ✓

3 marks

Gradient
- Negative gradient indicates a downward sloping line. This checks with the graph.
- Alternative calculation is:

$\dfrac{40-22}{2-8} = \dfrac{18}{-6} = -3$

SUCCESS GUIDE: pages 36, 37

Substitution
- $(a, b) = (8, 22)$ gives the same result.

Simplification
- Alternative versions are:
$y + 3x = 46$ or $y + 3x - 46 = 0$
SUCCESS GUIDE: page 55

Q11. (b) The x-axis shows the number of operators

So let $x = 6$

Then $y = -3 \times 6 + 46$

$= -18 + 46 = 28$ ✓

He might expect 28 complaints

1 mark

Substitution
- $x = 6$ is the substitution for a given number of operators not $y = 6$. This would be the number of daily complaints.

SUCCESS GUIDE: page 57

Q12. Area = length × breadth

$\Rightarrow 10\sqrt{2} = \text{length} \times \sqrt{10}$ ✓

$\Rightarrow \dfrac{10\sqrt{2}}{\sqrt{10}} = \text{length}$

$\Rightarrow \text{length} = \dfrac{10\sqrt{2} \times \sqrt{10}}{\sqrt{10} \times \sqrt{10}} = \dfrac{10\sqrt{20}}{10}$ ✓

$= \sqrt{20} = \sqrt{4 \times 5} = 2\sqrt{5}$ cm ✓

3 marks

Equation
- This 1st mark is awarded for correct substitution of Area $= 10\sqrt{2}$ and breadth $= \sqrt{10}$ into the Area formula.

Solution
- Solving the equation and 'rationalising the denominator' gains you the 2nd mark.
- Divide both sides of the equation by $\sqrt{10}$
- 'Rationalising the denominator' means getting rid of the square root sign on the bottom of the fraction. Remember in general $\sqrt{a} \times \sqrt{a} = a$.

Simplification
- Remove any 'square factors', in this case 4, from under the square root sign: $\sqrt{4} = 2$.

SUCCESS GUIDE: pages 9, 10

Q13. (a) (i)

Median $= 23$ ✓

(Q_2)

1 mark

Median
- The symbol Q_2 is used for the median.
- For 21 ages: (1st... 10th) (11th) (12th... 21st). So when placed in increasing order the 11th age gives the median age for this data, i.e. 23.

Q13. (a) (ii)

Lower Quartile $= \dfrac{21 + 22}{2} = 21 \cdot 5$ ✓

(Q_1)

1 mark

Lower Quartile
- The symbol Q_1 is used for the lower quartile
- This is the median of the lower half of the ages, i.e. the median of the 1st ten of the ages in increasing order: (1st... 5th) (6th... 10th). It is the mean of the 5th and 6th ages, i.e. the mean of 21 and 22.

Q13. (a) (iii)

Upper Quartile $= \dfrac{23 + 24}{2} = 23 \cdot 5$ ✓

(Q_3)

1 mark

Upper Quartile
- The Symbol Q_3 is used for the upper quartile
- This is the median of the upper half of the ages i.e. the median of the 2nd group of 10 ages in increasing order: (12th... 16th) (17th... 21st). It is the mean of the 16th and 17th ages, i.e. the mean of 23 and 24.

SUCCESS GUIDE: page 127.

Q13. (b) For oil rig Alpha:
Semi-interquartile range
$$= \frac{23.5 - 21.5}{2} = 1 \qquad \checkmark$$
The age distribution is more spread out (about the median age) on oil rig Beta compared to that on oil rig Alpha since the semi-interquartile range of 3.5 for Beta is greater than the 1 for Alpha. $\qquad \checkmark$

2 marks

Semi-interquartile range
- You need to calculate the statistic: Semi-interquartile range $= \frac{Q_3 - Q_1}{2}$ for oil rig Alpha and then compare it to the same statistic for oil rig Beta.

Comparison
- It is important you back up any comparison statement with the relevant statistics. In this case the semi-interquartile range: the greater the range the more spread out the data is around the median.
SUCCESS GUIDE: pages 128, 129

Q14. (a) Let u be the charge on an 'up' quark and let d be the charge on a 'down' quark:
$$2u + d = 1 \qquad \checkmark$$

1 mark

Setting up an equation
- It is important to be clear what the letters you use are representing. Here it is the value of the electric charge.

Q14. (b) $u + 2d = 0 \qquad \checkmark$

1 mark

Setting up an equation
- 'Algebraic' mentioned in a question means that algebra should be used, for example, equation solving, letter substitution, etc.

Q14. (c) Solve simultaneous equations
$$\left. \begin{array}{l} 2u + d = 1 \\ u + 2d = 0 \end{array} \right\} \begin{array}{l} \times 2 \rightarrow \\ \end{array} \quad \begin{array}{l} 4u + 2d = 2 \\ u + 2d = 0 \end{array} \checkmark$$
$$\text{subtract } 3u \qquad = 2$$
$$\Rightarrow u = \frac{2}{3} \qquad \checkmark$$

2 marks

Strategy
- Two equations with two unknowns (letters) indicates simultaneous equations need to be solved to find the values of the unknowns.

Method
- Line up the equations: in this case 'u's are lined up and 'd's are lined up.
- The aim is always to eliminate one of the unknowns, in this case d.

Calculation
- Divide both sides by 3.
SUCCESS GUIDE: page 61

Q15. (a) $9^{-\frac{1}{2}} = \dfrac{1}{9^{\frac{1}{2}}} \qquad \checkmark$
$$= \frac{1}{\sqrt{9}} = \frac{1}{3} \qquad \checkmark$$

2 marks

Negative indices
- $a^{-n} = \dfrac{1}{a^n}$ is the general Index Law used

Fractional indices
- $a^{\frac{m}{n}} = \left(\sqrt[n]{a} \right)^m$ is the general Index Law used.
This may help:
$$a^{\frac{m}{n}} \begin{array}{l} \leftarrow \text{power} \\ \leftarrow \text{root} \end{array} \qquad \text{so } a^{\frac{1}{2}} \begin{array}{l} \leftarrow \text{power 1} \\ \leftarrow \text{square root} \end{array}$$
So $9^{\frac{1}{2}}$ means the square root of 9 to the power 1.
SUCCESS GUIDE: pages 17, 18

Q15. (b) $\sqrt{t} \times t^2$

$= t^{\frac{1}{2}} \times t^2$ ✓

$= t^{\frac{1}{2}+2} = t^{\frac{5}{2}}$ ✓

2 marks

Simplifying indices
- $a^m \times a^n = a^{m+n}$ is the Index law used.

- Note $t^{2\frac{1}{2}}$ is not acceptable.

SUCCESS GUIDE: page 17

Q16. (a) (i) ✓

$A = (2, -9)$ ✓

2 marks

x-coordinate
- The least value of $(x-2)^2 - 9$ is $0^2 - 9$ and this occurs when $x = 2$. Remember something squared must be zero or positive, never a negative value.

y-coordinate
- when $x = 2, y = (2-2)^2 - 9 = 0^2 - 9 = -9$

SUCCESS GUIDE: page 71

Q16. (a) (ii)

$B = (2, 9)$ ✓

1 mark

Turning point
- By symmetry, point B is the reflection of point A in the x-axis so the x-coordinate will be the same (2) and the y-coordinate positive 9.

Q16. (b) The other parabola is a reflection of $y = (x-2)^2 - 9$ in the x-axis and so has the equation: ✓

$y = -[(x-2)^2 - 9]$

$\Rightarrow y = -(x-2)^2 + 9$

$\Rightarrow y = 9 - (x-2)^2$ ✓

2 marks

Reflection in x-axis
- If graph $y = $ (expression in x) is reflected in the x-axis then the equation of the image or new graph is given by $y = -$(expression in x). In this case you require the negative of $(x-2)^2 - 9$.

Equation
- It is better if you simplify $-[(x-2)^2 - 9]$. In a simpler example $-(a-b) = -a + b$. Following this pattern gives $-(x-2)^2 + 9$.
- Notice that, for example, $-2 + 3 = 3 - 2$ so $-(x-2)^2 + 9$ can be written $9 - (x-2)^2$. This is the more usual form for this expression.

SUCCESS GUIDE: pages 71, 75

Exam B: Paper 2

Q1. The annual increase factor is ✓
1·012. The factor is applied three
times (2007−2010).
Total amount in 2010 ✓
$= 7\cdot2 \times 1\cdot012 \times 1\cdot012 \times 1\cdot012$
$= 7\cdot46...$ ✓
$\doteqdot 7\cdot5$ million tonnes ✓
(correct to 2 significant figures).

4 marks

Multiplication factor
- An increase of 1·2% is the equivalent of calculating 101·2%. Multiplying by 1·012 calculates this amount.

Time calculation
- From 2007 to 2010 involves 3 complete years. Since each year's increase is 1·2 %, three multiplications by 1·012 are necessary.
- A shorter version is $7\cdot2 \times 1\cdot012^3$.

Calculation
- $\boxed{\wedge}\ \boxed{3}$ These keys mean 'raised to the power 3' or 'cubed'.

Rounding
- When accuracy is mentioned in the question there will be 1 mark allocated for the correct rounding.
SUCCESS GUIDE: pages 118, 119

Q2. (a) Mean
$$= \frac{44 + 47 + 38 + 97 + 40 + 52}{6}$$
$$= £53 \quad ✓$$

1 mark

Mean
- $\text{Mean} = \dfrac{\text{total of the numbers}}{\text{the number of numbers}}$
SUCCESS GUIDE: page 128

Q2. (b) There are 6 pieces of data
so $n = 6$
From part (a) $\bar{x} = 53$

x	$x-\bar{x}$	$(x-\bar{x})^2$
44	−9	81
47	−6	36
38	−15	225
97	44	1936
40	−13	169
52	−1	1

$\Sigma(x-\bar{x})^2 = 2448$

$s = \sqrt{\dfrac{\Sigma(x-\bar{x})^2}{n-1}}$

$= \sqrt{\dfrac{2448}{5}}$ ✓

$= \sqrt{489\cdot6}$

$= 22\cdot126...$

$\doteqdot \underline{£22\cdot13}$ ✓

(to 2 decimal places)

2 marks

Standard deviation
- You should always double check your calculations and if possible confirm your answer using the STAT mode on your calculator.
- It is important that all your working is shown. You will not gain the marks by calculating the standard deviation on your calculator and just writing the answer down.

Calculation
- Any correct rounding will gain the mark so long as your calculation is correct.
- Using the other formula gives $\Sigma x = 318$, $\Sigma x^2 = 19302$ and n = 6 resulting in $s \doteqdot 22\cdot13$
SUCCESS GUIDE: pages 130, 131

Q2. (c) There was less variation in the amounts spent on a weekday:
$(s = 8\cdot4)$ than on a Saturday
$(s = 22\cdot9,$ greater than $8\cdot4)$. ✓

1 mark

Comparison using statistics
- The standard deviation, s, measures the variation of the data about the mean. A greater value of s means more variation; a lower value means less variation.

Q3. Angle CEL $= 320° - 270°$
$\qquad\qquad = 50°$ ✓

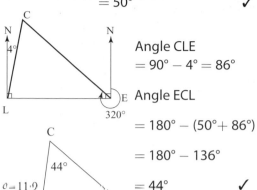

Angle CLE
$= 90° - 4° = 86°$

Angle ECL
$= 180° - (50° + 86°)$
$= 180° - 136°$
$= 44°$ ✓

Use the sine rule: $\dfrac{c}{\sin C} = \dfrac{e}{\sin E}$

so $\dfrac{c}{\sin 44°} = \dfrac{11\cdot9}{\sin 50°} \Rightarrow c = \dfrac{11\cdot9 \times \sin 44°}{\sin 50°}$ ✓

$= 10\cdot79\ldots$

So required distance $\doteqdot 10\cdot8$ km ✓
(correct to 1 decimal place)

Diagram
- In 'Bearing' questions diagrams are essential.
- 'Due west' gives the clue that the Leven to Elie line LE is at right-angles to the North line.
- Bearings are always measured clockwise from the North line.

SUCCESS GUIDE: page 110

Angle sum in triangle
- To use the Sine Rule the 3rd angle must be calculated.

Sine rule
- Multiply both sides by sin 44° to calculate c so that sin 44° will appear at the top of the fraction.

Calculation
- Always check that your answer, 10·8 km, appears reasonable in the given context. In this case it is comparable to 11·9 km, the other length, and so seems reasonable.

SUCCESS GUIDE: page 111

4 marks

Q4. Find the two volumes ✓
The wok
Volume of a sphere $= \dfrac{4}{3}\pi r^3$

So volume of a hemisphere
$= \dfrac{1}{2} \times \dfrac{4}{3}\pi r^3 = \dfrac{2}{3}\pi r^3$

In this case, $r = \dfrac{31}{2} = 15\cdot5$ cm ✓

So volume $= \dfrac{2}{3} \times \pi \times 15\cdot5^3$

$= 7799\cdot26\ldots \doteqdot 7800$ cm³
$\qquad\qquad$ (to 3 sig. figs)

The pan
Volume of a cylinder $= \pi r^2 h$
In this case, $r = 25$ cm and
$h = 3\cdot5$ cm
So volume $= \pi \times 25^2 \times 3\cdot5$ ✓
$= 6872\cdot23\ldots \doteqdot 6900$ cm³
$\qquad\qquad$ (to 3 significant figures) ✓
So the wok has the larger capacity
by approximately 900 cm³. ✓

Strategy
- Capacity means volume. You will have to calculate the volume of each type of pan and compare these volumes.

1st substitution
- The formula $V = \dfrac{4}{3}\pi r^3$ is given to you on the formulae page during your exam.
- In this case you have only half a sphere so use $\dfrac{1}{2}$ of $\dfrac{4}{3}\pi r^3$ i.e. $\dfrac{2}{3}\pi r^3$ with r being half of the diameter of 31 cm, i.e. 15·5 cm.

2nd substitution
- The formula $V = \pi r^2 h$ is not given on your formulae sheet.
- In this case substitute $r = 25$ and $h = 3\cdot5$.

Volumes
- Both volumes correct and you gain this mark.

Statement
- You must clearly justify your conclusion using the values for the volume. Finding the difference and stating this shows clearly that you understand the results you calculated.

SUCCESS GUIDE: pages 40, 41

5 marks

Q5. (a) $\overrightarrow{PA} = \overrightarrow{PD} + \overrightarrow{DA}$

represents $v + (-u) = v - u$ ✓

1 mark

Expression

- $\overrightarrow{AD}$ represents u. Then $\overrightarrow{DA}$ represents $-u$.

Q5. (b) $\overrightarrow{BD} = \overrightarrow{BP} + \overrightarrow{PD}$

now $\overrightarrow{BP} = \overrightarrow{PA}$, which represents

$v - u$ from part (a)

So $\overrightarrow{BD}$ represents

$\underset{(\overrightarrow{BP})}{v - u} + \underset{(\overrightarrow{PD})}{v} = 2v - u$ ✓

1 mark

Expression

- Two directed line segments like $\overrightarrow{PA}$ and $\overrightarrow{BP}$ with the same length and direction will represent the same vector.

SUCCESS GUIDE: pages 112, 113

Q6. Distance in km $= \dfrac{4 \cdot 014 \times 10^{16}}{10^3}$

$= 4 \cdot 014 \times 10^{13}$

Speed of light $= 9 \cdot 461 \times 10^{12}$ km/yr

Time $= \dfrac{\text{Distance}}{\text{Speed}} = \dfrac{4 \cdot 014 \times 10^{13}}{9 \cdot 461 \times 10^{12}}$ years ✓

$= 4 \cdot 2426 \dots$

$\doteqdot 4 \cdot 24$ years ✓

(to 2 decimal places)

2 marks

Correct method

- You should be using $T = \dfrac{D}{S}$

- Be careful that units match when using the formula. You need to change both measurements to metres or both to kilometres.

Calculation

- Use the EE or EXP keys on your calculator to enter the values; e.g. 4.014 EXP 13 for 4.014×10^{13}
- Alternatively calculate: $\dfrac{4.014}{9.461} \times 10^1$

SUCCESS GUIDE: page 19

Q7. Use the Cosine Rule:

$a^2 = b^2 + c^2 - 2bc \cos A$ ✓

$\Rightarrow a^2 = 12^2 + 10^2 - 2 \times 12 \times 10 \times \cos 10°$ ✓

$\Rightarrow a^2 = 7 \cdot 646 \dots$ ✓

$\Rightarrow a = \sqrt{7 \cdot 646 \dots} = 2 \cdot 765 \dots$

$\Rightarrow BC \doteqdot 2 \cdot 77$ km ✓

(to 3 significant figures)

3 marks

Substitution

- $b = 12$, $c = 10$ and $A = 10°$

Calculation

- Correct calculation of a^2 earns this mark.

Answer

- Any correct approximation earns this mark.

SUCCESS GUIDE: page 108

Q8. (a) $3d + t = 15$

$\Rightarrow 3d = -t + 15$

$\Rightarrow d = -\frac{1}{3}t + 5$ ✓

so gradient $= -\frac{1}{3}$ ✓

2 marks

Rearrangement

- Rearrange into the form $y = mx + c$ but in this case use: d instead of y

  and t instead of x.

Gradient

- Compare: $y = -\frac{1}{3}x + 5 \Rightarrow$ gradient $= -\frac{1}{3}$

SUCCESS GUIDE: page 52

Q8. (b) $A(0,5)$ ✓

1 mark

Coordinates
- Brackets must be used so $d = 5$ will not earn this mark.

Q9.

$$\frac{1}{(\sqrt[3]{a})^2} = \frac{1}{a^{\frac{2}{3}}}$$ ✓

$$= a^{-\frac{2}{3}}$$ ✓

2 marks

1st Index Law
- You are using $(\sqrt[n]{a})^m = a^{\frac{m}{n}}$ with $n = 3$ and $m = 2$.

2nd Index Law
- You are using $\frac{1}{x^n} = x^{-n}$ with $n = \frac{2}{3}$.

SUCCESS GUIDE: pages 17–18

Q10.

$$\frac{2-a}{a^2} + \frac{1}{a}$$ ✓

$$= \frac{2-a}{a^2} + \frac{a}{a^2}$$ ✓

$$= \frac{2-a+a}{a^2} = \frac{2}{a^2}$$ ✓

3 marks

Common denominator
- To add, for example, $\frac{1}{2} + \frac{1}{3}$ you need to change the fractions to $\frac{3}{6} + \frac{2}{6}$ so that they have a 'common denominator' of 6. In this case $\frac{2-a}{a^2} + \frac{1}{a}$ cannot be added until their denominators are the same. 'a^2' is the simplest such 'common denominator'.

Equivalent fraction
- $\frac{1}{a} = \frac{1 \times a}{a \times a} = \frac{a}{a^2}$. You are always allowed to multiply (or divide) the top and bottom of a fraction by the same number or expression.

Simplify
- $\frac{3}{6} + \frac{2}{6} = \frac{3+2}{6}$: the numerators are added once the denominators are the same. In this case the numerators are $2 - a$ and a giving $2 - a + a = 2$.

SUCCESS GUIDE: page 33

Q11.

$$P = \sqrt{9 - Q^2}$$ ✓

$$\Rightarrow P^2 = 9 - Q^2$$

$$\Rightarrow Q^2 = 9 - P^2$$ ✓

$$\Rightarrow Q = \sqrt{9 - P^2}$$ ✓

3 marks

1st step
- Your aim is to isolate the letter Q. The first step is to remove the square root. This is done by squaring both sides.

2nd step
- Add Q^2 to both sides and substract P^2 from both sides. This isolates Q^2.

Final step
- To remove the 'squaring' you must 'square root' both sides.
- Technically you should write $\pm\sqrt{9 - P^2}$ but you will not be penalised for not doing this!

SUCCESS GUIDE: pages 64–66

Q12. For x-axis intercepts set $y = 0$

So $5\sin x° - 1 = 0$ ✓

$\Rightarrow 5\sin x° = 1$

$\Rightarrow \sin x° = \dfrac{1}{5}$ ✓

Since this is positive $x°$ is in the 1st and 2nd quadrants

The 1st quadrant angle is given by

$\sin^{-1}\left(\dfrac{1}{5}\right) = \sin^{-1}(0\cdot2) = 11\cdot5°$ ✓

The 2nd quadrant angle is

$180° - 11\cdot5° = 168\cdot5°$

So the required x-coordinate values are $11\cdot5$ and $168\cdot5$. ✓

4 marks

Equation
- This mark is for obtaining the correct equation.

Rearrangement
- The correct rearrangement of the equation gives you a value for $\cos x°$.

x-intercept
- Check that 11·5 is 'sensible' by looking at the graph.

x-intercept
- use $\dfrac{S \mid A}{T \mid C}$ as a memory aid:

sin $x°$ is positive in the 1st *and* 2nd quadrants.
SUCCESS GUIDE: pages 96, 97

Q13.

$q = 2\cdot5$, P, $r = 3\cdot4$, R, Q

Area of triangle $= \dfrac{1}{2}qr\sin P$ ✓

so $4\cdot2 = \dfrac{1}{2} \times 2\cdot5 \times 3\cdot4 \times \sin P$

$\Rightarrow 4\cdot2 = 4\cdot25 \times \sin P$

$\Rightarrow \dfrac{4\cdot2}{4\cdot25} = \sin P$

so $\sin P = 0\cdot9882...$ ✓

$\Rightarrow$ angle $P = \sin^{-1}(0\cdot9882...)$

$= 81\cdot2...°$

so angle $P \doteqdot 81°$ ✓

(to 2 significant figures)

3 marks

Strategy
- Normally the formula 'Area $= \dfrac{1}{2}ab\sin C$' is used to find the area but in this case the area is given so an equation is set up and solved first for sin P.

Calculation
- The aim is to find a value for sin P.

Angle
- Use $\boxed{\sin^{-1}}\,\boxed{\text{ans}}\,\boxed{\text{EXE}}$ on your calculator where 'ans' uses the previous answer, 0·9882..., on the calculator display.
- $\boxed{\text{EXE}}$ and $\boxed{=}$ are the same – it depends on your calculator.
SUCCESS GUIDE: pages 96, 104, 105

Q14. (a) Set $y = 0$ for x-axis intercepts

so $k(x - p)(x - q) = 0$

$\Rightarrow x - p = 0$ or $x - q = 0$

$\Rightarrow x = p$ or $x = q$

From the given graph this gives: ✓

$p = -2$ and $q = 6$ ✓

2 marks

Intercepts
- There is a correspondence between factors and x-axis intercepts:

Factor	Intercept
$x - p$	$(p, 0)$
$x - q$	$(q, 0)$

- Since $p = -2$ the factor $x - p = x - (-2) = x + 2$ so $x + 2$ as a factor gives $(-2, 0)$ as an x-axis intercept.
SUCCESS GUIDE: page 72

Q14. (b) The equation is $y = k(x+2)(x-6)$ and from the graph $(0, 6)$ lies on the parabola.
So substitute $x = 0$ and $y = 6$ into the equation of the parabola: ✓

$6 = k \times (0+2) \times (0-6)$

$\Rightarrow 6 = k \times 2 \times (-6)$

$\Rightarrow 6 = k \times (-12)$

$\Rightarrow k = \dfrac{6}{-12} = -\dfrac{1}{2}$ ✓

2 marks

Substitution
- If a point (a, b) lies on a graph then values $x = a$ and $y = b$ will satisfy the equation of that graph. Substituting these values into the equation gives an equation with only one unknown k in this case.

Calculation
- Some graphs with equation $y = k(x+2)(x-6)$ are shown. They all have x-intercepts $(-2, 0)$ and $(6, 0)$ and are all parabolas. Only one graph passes through $(0, 6)$:

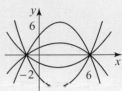

$y = -\dfrac{1}{2}(x+2)(x-6)$,

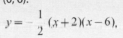

Q14. (c) The x value that gives the maximum is halfway between -2 and 6;

i.e. $x = \dfrac{-2+6}{2} = \dfrac{4}{2} = 2$ ✓

The equation is:

$y = -\dfrac{1}{2}(x+2)(x-6)$

Now substitute $x = 2$

so $y = -\dfrac{1}{2}(2+2)(2-6)$

$= -\dfrac{1}{2} \times 4 \times (-4) = 8$

The maximum turning point is $(2, 8)$ ✓

2 marks

x-value
- The graph is symmetrical with the axis of symmetry $x = 2$.
- Required value is the mean of -2 and 6.

y-value
- The question asks for the coordinates of a point, i.e. $(2, 8)$ not the separate values.
SUCCESS GUIDE: page 73

Q15. The smaller volume is required so a reduction scale factor is needed.

Length scale factor:

$\dfrac{6}{8} = \dfrac{3}{4} = 0.75$

Volume scale factor: 0.75^3 ✓
Required volume

$= 43 \times 0.75^3$

$= 18.14....$ ✓

$\doteqdot 18$ litres ✓
(to the nearest litre)

3 marks

Scale factor
- If the length scale factor is k then the area scale factor is k^2; and the volume scale factor is k^3.

Calculation
- For a reduction, the scale factor should lie between 0 and 1. You should check that your final answer is smaller than the given volume of 43 litres.

Rounding
- 1 mark is allocated for correct rounding to the nearest litre since the required accuracy is mentioned in the question.
SUCCESS GUIDE: pages 87, 88

Q16. AC is a radius
so AC = 2·5 cm

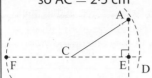

CD is also
a radius
so CD = 2·5 cm ✓

$ED = DF - EF = 5 - 4\cdot8 = 0\cdot2$ cm
and $CE = CD - ED$
$= 2\cdot5 - 0\cdot2 = 2\cdot3$ cm ✓
Use Pythagoras's Theorem in

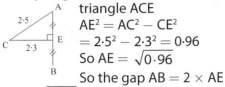

triangle ACE
$AE^2 = AC^2 - CE^2$
$= 2\cdot5^2 - 2\cdot3^2 = 0\cdot96$
So $AE = \sqrt{0\cdot96}$ ✓
So the gap $AB = 2 \times AE$
$= 2 \times \sqrt{0\cdot96}$
$= 1\cdot959...$
$\doteq 2\cdot0$ cm (to 1 decimal place) ✓

4 marks

Strategy
- The construction of triangle ACE and the use of Pythagoras's Theorem are the essential steps in this solution.

Use of radius
- All radii in a circle are equal. The diameter in this case is 5 cm so the radius $= \frac{1}{2} \times 5$ cm.

Calculation
- The symmetry of the diagram (CE is an axis of symmetry) means that the gap AB is double the length AE.

Answer
- The value 2·0 cm seems reasonable in this context. The value of $\sqrt{0\cdot96}$ should not be rounded before the final answer is reached.
- Any correct rounding is acceptable in the final answer as accuracy is not mentioned in the question.

SUCCESS GUIDE: pages 76, 85

Q17. (a) Using $S = \dfrac{D}{T}$

Average speed

$= \dfrac{180}{x}$ m.p.h. ✓

1 mark

D.S.T. formula
- Covering up S reveals $\frac{D}{T}$ as the formula for calculating average speed.

Q17. (b) Again, using $S = \dfrac{D}{T}$

"2 hours less" gives $x - 2$ hours

So Average speed $= \dfrac{60}{x-2}$ m.p.h. ✓

1 mark

Using algebra
- You should be aware that '2 less than x' translates to the algebraic expression $x - 2$
- Note that the distance has changed to 60 miles.

Q17. (c) The two average speeds are equal so:
$$\frac{180}{x} = \frac{60}{x-2}$$ ✓
$\Rightarrow 180(x-2) = 60x$
$\Rightarrow 180x - 360 = 60x$
$\Rightarrow 120x = 360$
$\Rightarrow x = 3$ ✓
1st stage took $x = 3$ hours
2nd stage took $x - 2 = 1$ hour
Total journey took 4 hours ✓

3 marks

Set up equation
- The strategy here is to equate the two expressions for the average speeds.
- You would still gain this mark even if you equated 'wrong' expressions.

Solving
- Multiply both sides of the equation by x and then by $x - 2$ (or 'cross-multiply')

Interpretation
- What does $x = 3$ mean? You must look back to x hours and $x - 2$ hours and substitute.

SUCCESS GUIDE: page 49

Answers to Exam C

Exam C: Paper 1

Q1. A (k, k^2) B $(1, k)$

So $m_{AB} = \dfrac{k^2 - k}{k - 1}$ ✓

$= \dfrac{k(k-1)}{k-1} = k$ ✓ ✓

3 marks

Gradient
- Use the formula
$$m = \frac{y_2 - y_1}{x_2 - x_1}$$
SUCCESS GUIDE: page 36.

Factorisation
- k is a common factor.

Cancellation
- You should make clear in your solution that the factor $k - 1$ cancels.
SUCCESS GUIDE: pages 29, 30.

Q2. $\dfrac{2}{3} \div 1\dfrac{1}{3}$

$= \dfrac{2}{3} \div \dfrac{4}{3} = \dfrac{\frac{2}{3}(\times 3)}{\frac{4}{3}(\times 3)}$ ✓

$= \dfrac{2}{4} = \dfrac{1}{2}$ ✓

2 marks

Division of Fractions
- Change mixed fractions to 'top-heavy'
- An alternative method to the one given is to 'invert then multiply'.
$$\text{So } \frac{2}{3} \div \frac{4}{3} = \frac{2}{3} \times \frac{3}{4} = \frac{6}{12} = \frac{1}{2}$$

Calculation
- Remember to 'cancel down' if possible.
SUCCESS GUIDE: page 122

Q3. $f(x) = x(2 - x)$

$\Rightarrow f(-1) = -1 \times (2 - (-1))$ ✓

$= -1 \times (2 + 1)$

$= -1 \times 3 = -3$ ✓

2 marks

Substitution
- for $f(-1)$, every occurrence of x is replaced by the value -1. No further calculation is needed to gain the 1st mark.

Calculation
- The 2nd mark here is for the subsequent calculation after the substitution.
- Remember: brackets first, and that subtracting a negative is the same as adding the positive value.
SUCCESS GUIDE: page 47

Q4. $3 + 2x < 4(x + 1)$

$\Rightarrow 3 + 2x < 4x + 4$ ✓

(now subtract 4 from each side)

$\Rightarrow -1 + 2x < 4x$

(now subtract $2x$ from each side)

$\Rightarrow -1 < 2x$ ✓

$\Rightarrow \dfrac{-1}{2} < x$

so $x > -\dfrac{1}{2}$ ✓

3 marks

Brackets
- Both terms are multiplied by 4. A common mistake here would be $4x + 1$.

Simplification
- The 'balancing process' aims to get letters on one side of the inequality sign and numbers on the other side.
- Best to avoid $-2x < 1$. The next step here would be divide by -2 and 'swap the sign round': $x > \dfrac{1}{-2}$

Answer
- Notice if $a < b$ then $b > a$
SUCCESS GUIDE: page 50

Q5. (a) $9y^2 - 4$

$= (3y)^2 - 2^2$

$= (3y - 2)(3y + 2)$ ✓

1 mark

Difference of squares
- The pattern is: $A^2 - B^2 = (A - B)(A + B)$.
- You should always check that your answer multiplies out giving the original expression (use 'FOIL').
SUCCESS GUIDE: page 27

Q5. (b) $\dfrac{9y^2 - 4}{15y - 10}$

$= \dfrac{(3y - 2)(3y + 2)}{5(3y - 2)}$ ✓

$= \dfrac{3y + 2}{5}$ ✓

2 marks

Factorising
- When simplifying algebraic fractions the 1st step is to factorise both expressions.
- 'Hence' appearing in a question means you should use the answer that you got in the previous part of the question: in this case $9y^2 - 4 = (3y - 2)(3y + 2)$.

Cancelling
- Any factor, e.g. $3y - 2$, that appears on the top and the bottom can be cancelled.
SUCCESS GUIDE: pages 29, 30.

Q6. $P = \dfrac{W}{4A}$

$(\times A)\ (\times A)$

$\Rightarrow PA = \dfrac{W}{4}$ ✓

$(\div P)\ (\div P)$

$\Rightarrow A = \dfrac{W}{4P}$ ✓

2 marks

1st rearrangement
- The subject of the formula, A, in this case appears on the bottom of the fraction, i.e. W is divided by A. The inverse process is multiplication so multiply both side of the formula by A.

2nd rearrangement
- A is multiplied by P. The inverse process is division so divide both sides by P.
SUCCESS GUIDE: page 64

Q7. Use the Cosine rule

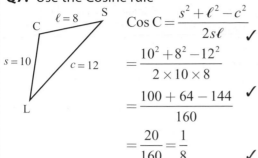

$$\text{Cos}\, C = \frac{s^2 + \ell^2 - c^2}{2s\ell} \quad \checkmark$$

$$= \frac{10^2 + 8^2 - 12^2}{2 \times 10 \times 8}$$

$$= \frac{100 + 64 - 144}{160} \quad \checkmark$$

$$= \frac{20}{160} = \frac{1}{8} \quad \checkmark$$

3 marks

Cosine formula
- The formula as given in the exam is:
$$\cos A = \frac{b^2 + c^2 - a^2}{2bc}$$
and you have to be able to adapt this to the particular letters used in the example.

Substitution
- Be very careful with the order of the letters. Swapping values, e.g. ℓ and c, will give you the wrong answer.

Calculation
- It is important to show all your working since the answer, $\frac{1}{8}$, is given.

SUCCESS GUIDE. page 109

Q8. $(2 + x)(3 - 2x - x^2)$

$= 2(3 - 2x - x^2) + x(3 - 2x - x^2) \quad \checkmark$

$= 6 - 4x - 2x^2 + 3x - 2x^2 - x^3 \quad \checkmark$

$= 6 - x - 4x^2 - x^3 \quad \checkmark$

3 marks

Start the expansion
- An alternative strategy is:
$3(2 + x) - 2x(2 + x) - x^2(2 + x)$

Complete the expansion
- All brackets should be removed.

Simplification
- Gather the like terms and simplify.
- $-x^3 - 4x^2 - x + 6$ is equally valid.

SUCCESS GUIDE: page 24

Q9. $f(x) = 3x^2 - x - 5$

compare $ax^2 + bx + c$

so $a = 3$, $b = -1$ and $c = -5$.

Discriminant $= b^2 - 4ac$

$= (-1)^2 - 4 \times 3 \times (-5)$

$= 1 + 60 = 61 > 0 \quad \checkmark$

Since the Discriminant > 0

there are two real and distinct roots. $\checkmark$

2 marks

Discriminant
- Most mistakes involve errors with the negative number calculations:
notice: $(-1)^2 = -1 \times (-1)$ is positive
and $-4 \times 3 \times (-5)$ is also positive.

Statement
- You must make a clear and unambiguous statement about the roots.

SUCCESS GUIDE: page 74

Q10. Use Area $= \frac{1}{2} ab \sin C$

with $a = 2$, $b = 5$ and $\sin C = \frac{7}{10} \quad \checkmark$

so Area $= \frac{1}{2} \times 2 \times 5 \times \frac{7}{10} = \frac{7}{2} = 3 \cdot 5\, \text{cm}^2 \quad \checkmark$

2 marks

Substitution
- You will gain this mark for correct substitution in the formula.

Calculation
- You have no calculator in this paper so cancellation is useful:

$$\frac{1}{\underset{1}{2}} \times \frac{\overset{1}{2}}{1} \times \frac{\overset{1}{5}}{1} \times \frac{7}{\underset{2}{10}} = \frac{7}{2}$$

SUCCESS GUIDE: page 104

Q11. (a) $a = -2$ ✓

1 mark

Value of a
- For $y = (x - 2)^2$ the graph $y = x^2$ is shifted 2 units to the right parallel to the x-axis.

Q11. (b) For the point $(3, -2)$

$x = 3$ and $y = -3$ so these values are substituted into the equation:

$$y = (x - 2)^2 + b$$
$$\Rightarrow -2 = (3 - 2)^2 + b \quad ✓$$
$$\Rightarrow -2 = 1 + b$$
$$\Rightarrow b = -3 \quad ✓$$

2 marks

Substitution
- The values of the coordinates will satisfy the equation since the point lies on the parabola.

Value of b
- Notice that the graph of $y = (x - 2)^2 - 3$ can be obtained from the graph $y = x^2$ by a shift of 2 units to the right parallel to the x-axis followed by a shift down of units parallel to the y-axis.

SUCCESS GUIDE: pages 70–71

Q12. (a)

M (4, 4, 0) ✓

1 mark

Coordinates
- The square base has side length 8 units.
- M has height (z-coordinate) of zero.

Q12. (b)

D (4, 4, 8) ✓

1 mark

Coordinates
- D has height 8 units, the same as the side of the square base.

SUCCESS GUIDE: page 92

Q12. (c) ✓

$$\left\| \begin{pmatrix} -4 \\ 4 \\ 8 \end{pmatrix} \right\| = \sqrt{(-4)^2 + 4^2 + 8^2}$$

$$= \sqrt{16 + 16 + 64} \quad ✓$$

$$= \sqrt{96} = \sqrt{16 \times 6} = 4\sqrt{6}$$

2 marks

Magnitude
- Use
$$\left\| \begin{pmatrix} a \\ b \\ c \end{pmatrix} \right\| = \sqrt{a^2 + b^2 + c^2}$$

Simplification
- Notice $16 + 16 + 64 = 16(1 + 1 + 4)$
$$= 16 \times 6$$

SUCCESS GUIDE: pages 116, 117

Q13. Compare $x^2 - 5x + \dfrac{1}{4}$

With $\left(x - \dfrac{5}{2}\right)^2 = x^2 - 5x + \dfrac{25}{4}$

so $x^2 - 5x + \dfrac{1}{4}$ ✓

$$= \left(x - \frac{5}{2}\right)^2 - \frac{25}{4} + \frac{1}{4}$$

$$= \left(x - \frac{5}{2}\right)^2 - \frac{24}{4} = \left(x - \frac{5}{2}\right)^2 - 6 \quad ✓$$

2 marks

Value of p
- This value is half the coefficient -5 of x.

Value of q
- $\dfrac{-5}{2} \times \left(\dfrac{-5}{2}\right) = \dfrac{5 \times 5}{2 \times 2} = \dfrac{25}{4}$ (positive)

- $\dfrac{-25}{4} + \dfrac{1}{4} = \dfrac{-25 + 1}{4} = \dfrac{-24}{4} = -6$

SUCCESS GUIDE: page 28

Q14. $f(x) = g(x)$

$\Rightarrow 4 + 3x - x^2 = 10 - 2x$ ✓

$\Rightarrow 0 = 10 - 2x - 4 - 3x + x^2$

$\Rightarrow x^2 - 5x + 6 = 0$ ✓

$\Rightarrow (x - 2)(x - 3) = 0$

$\Rightarrow x - 2 = 0$ or $x - 3 = 0$

$\Rightarrow x = 2$ or $x = 3$ ✓

3 marks

Set up equation
- $f(x)$ is replaced by $4 + 3x - x^2$ and $g(x)$ is replaced by $10 - 2x$.

Standard form
- When you recognise a quadratic equation then rearrange it to the form:
$ax^2 + bx + c = 0$.

Solving quadratic equation
- Solving a quadratic equation will not require 'the formula' unless the question asks for the roots 'to 1 decimal place' or similar.
SUCCESS GUIDE: pages 67, 68

Q15. $\sqrt{(\sqrt{18})^2 + (\sqrt{6})^2}$

$= \sqrt{18 + 6} = \sqrt{24}$ ✓

$= \sqrt{4 \times 6} = \sqrt{4} \times \sqrt{6}$

$= 2 \times \sqrt{6} = 2\sqrt{6}$ ✓

2 marks

1st simplification
- $(\sqrt{18})^2 = \sqrt{18} \times \sqrt{18} = 18$. In general you should know: $\sqrt{a} \times \sqrt{a} = a$.

2nd simplification
- $\sqrt{24}$ is simplified when all 'square factors' have been removed: in this case 4.
SUCCESS GUIDE: page 9

Q16. $x^3 \times (x^{-1})^{-2}$

$= x^3 \times x^{-1 \times (-2)} = x^3 \times x^2$ ✓

$= x^{3+2} = x^5$ ✓

2 marks

Laws of indices
- The 1st law used here is $(a^m)^n = a^{mn}$
- The 2nd law used is $a^m \times a^n = a^{m+n}$
SUCCESS GUIDE: page 17

Q17. Gradient $= \dfrac{24 - 20}{16} = \dfrac{4}{16} = \dfrac{1}{4}$ ✓

The intercept is (0, 20) ✓

Equation is $L = \dfrac{1}{4}W + 20$ ✓

3 marks

Gradient
- Gradient $= \dfrac{\text{distance up (or down)}}{\text{distance along}}$

- Take care when reading the scales as they are not the same on the two axes.

'$y = mx + c$'
- The equation follows the pattern:

$y = mx + c$

this is L   gradient $= \dfrac{1}{2}$ this is W   (0, 20) is the intercept

SUCCESS GUIDE: page 57

Q18. (a) Let k euros be the cost of 1 kg of Kenyan and r euros be the cost of 1 kg of Rwandan

so $2k + 3r = 6 \cdot 1$ ✓

1 mark

1st equation
- 'Algebraic' refers to using letters.
- It is important to state clearly what each letter stands for: in this case the cost of 1 kg.

Q18. (b)

$3k + 2r = 5 \cdot 9$ ✓

1 mark

2nd equation
- Notice that there are no units appearing in the equations.
- When simultaneous equations are anticipated it is useful to write the equations in the form: $ax + by = c$ where a, b and c are the given values, e.g. 2, 3, 6·1 or 3, 2, 5·9, and x and y are the variables, in this case k and r.

Q18. (c) Solve simultaneous equations:

$$\begin{aligned} 2k + 3r = 6 \cdot 1 &\ \big] \times 3 \rightarrow 6k + 9r = 18 \cdot 3 \ ✓ \\ 3k + 2r = 5 \cdot 9 &\ \big] \times 2 \rightarrow \underline{6k + 4r = 11 \cdot 8} \end{aligned}$$

Subtract: $5r = 6 \cdot 5$

$\Rightarrow\quad r = 1 \cdot 3$ ✓

Substitute $r = 1 \cdot 3$ in $2k + 3r = 6 \cdot 1$

$\Rightarrow 2k + 3 \times 1 \cdot 3 = 6 \cdot 1$

$\Rightarrow 2k + 3 \cdot 9 = 6 \cdot 1$

$\Rightarrow 2k = 2 \cdot 2 \Rightarrow k = 1 \cdot 1$ ✓

So 1 kg of Kenyan costs 1·10 euros and 1kg of Rwandan costs 1·30 euros

So 4 kg of Kenyan and 1 kg of Rwandan costs

$4 \times 1 \cdot 10 + 1 \times 1 \cdot 30$

$= 5 \cdot 70$ euros

5 kg of Blend C costs 5·70 euros ✓

4 marks

Method
- The 'simultaneous equation' strategy will be rewarded with a 'strategy mark'.

1st variable
- A further mark is awarded for the correct calculation of one of the variables, either k or r.
- In the working shown 'k' has been eliminated. 'r' could have been eliminated by multiplying the top equation by 2 and the bottom equation by 3 and then subtracting.

2nd variable
- Either of the two equations can be chosen for substitution of the 1st value to calculate the 2nd value.

Answer
- The question did not ask for just the values of k and r!

SUCCESS GUIDE: pages 61–63

Exam C: Paper 2

Q1. Number of letters $= \dfrac{1 \cdot 05 \times 10^{-1}}{3 \times 10^{-8}}$ ✓

$= 3\,500\,000$

$= 3 \cdot 5 \times 10^6$ ✓

2 marks

Strategy
- This is a division; it is sometimes useful to try to solve a simpler similar problem to see how the solution is found:
- If each letter were 2 mm wide and the line was 6 mm long there would be 3 letters: 6 divided by 2 gives 3.

Calculation
- $1 \cdot 05 \times 10^{-1}$ is entered in the calculator as

  follows: $\boxed{1}\ \boxed{\cdot}\ \boxed{0}\ \boxed{5}\ \boxed{\times 10^x}\ \boxed{-}\ \boxed{1}$
- On your calculator, you may have $\boxed{EXP}$ or $\boxed{EE}$ not $\boxed{\times 10^x}$

SUCCESS GUIDE: pages 18, 19

Q2. It is now $106\frac{1}{2}$ % of the original cost so:

$106 \cdot 5\,\% \leftrightarrow £87 \cdot 82$ ✓

$1\,\% \leftrightarrow \dfrac{£87 \cdot 82}{106 \cdot 5}$

$100\,\% \leftrightarrow \dfrac{£87 \cdot 82}{106 \cdot 5} \times 100$ ✓

$= £82 \cdot 46$ ✓

3 marks

Strategy
- The 'original' price has to be found. It will be 100 %. This is now a 'proportion' problem: find 1 % then find 100 %.

Calculation
- It is sometimes best not to calculate intermediate answers, e.g. £87·82 ÷ 106·5 but build up the calculation on paper and then do the calculation at the end. The solution does just that.

SUCCESS GUIDE: pages 119, 120

Q3. $\qquad 3x^2 - x - 5 = 0$

Compare $ax^2 + bx + c = 0$

$\Rightarrow a = 3, b = -1, c = -5$

$x = \dfrac{-b \pm \sqrt{b^2 - 4ac}}{2a}$

$= \dfrac{-(-1) \pm \sqrt{(-1)^2 - 4 \times 3 \times (-5)}}{2 \times 3}$ ✓

$= \dfrac{1 \pm \sqrt{(1 + 60)}}{6} = \dfrac{1 \pm \sqrt{61}}{6}$ ✓

so $x = \dfrac{1 + \sqrt{61}}{6}$ or $x = \dfrac{1 - \sqrt{61}}{6}$

$x = 1 \cdot 468 \ldots$ or $x = -1 \cdot 135 \ldots$ ✓

$x \doteqdot 1 \cdot 5 \ or \ x \doteqdot -1 \cdot 1$ ✓

(correct to one decimal place)

4 marks

Substitution
- Care should be taken over negative values.
- It is safer to state the values of a, b and c that you are going to use. Correct substitution into the formula will gain the 1st mark.

Simplification
- Before 'going decimal' it is clearer, and avoids mistakes, to separate the two solutions (roots) as is done in the 3rd last line.

Calculation
- Giving a clear indication of each decimal before it is rounded will allow you to gain the last mark for rounding even if your answers are wrong before rounding.

Rounding
- Always follow precisely any rounding requests.

SUCCESS GUIDE: pages 68, 69

Q4.

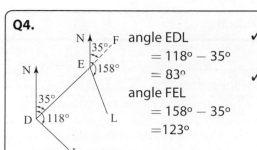

angle EDL ✓
$$= 118° - 35°$$
$$= 83°$$ ✓

angle FEL
$$= 158° - 35°$$
$$= 123°$$

So angle DEL
$$= 180° - 123° = 57°$$
angle ELD $= 180° - (83° + 57°)$
$$= 180° - 140° = 40°$$ ✓

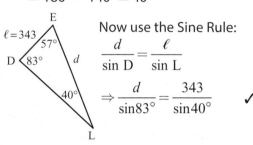

Now use the Sine Rule:
$$\frac{d}{\sin D} = \frac{\ell}{\sin L}$$
$$\Rightarrow \frac{d}{\sin 83°} = \frac{343}{\sin 40°}$$ ✓

$$\Rightarrow d = \frac{343 \times \sin 83°}{\sin 40°} = 529 \cdot 63\ldots$$
$$\doteqdot 530$$ ✓

The distance from Edinburgh to London is approximately 530 km to the nearest km.

5 marks

Strategy
- You need to calculate the three angles in triangle DEL and then use the Sine Rule.

1st angle
- Remember bearings are always measured clockwise from the North line.

2nd & 3rd angles
- Extending the line DE (EF in the diagram) allows you to move the information at angle D up to angle E (35° is a corresponding angle).
SUCCESS GUIDE: page 88

The Sine rule
- Substitution of the values into the Sine rule will gain you a mark.

Calculation
- Multiply both sides of the equation by sin 83°.
- Any correct rounding is acceptable.
SUCCESS GUIDE: pages 110, 111

Q5. Use $A = \pi r^2$

Outside circle diameter
$= 30$ cm $\Rightarrow$ radius $= 15$ cm

Inside circle diameter
$$= 30 - 2 \times 2\frac{1}{2}$$
$$= 25 \text{ cm}$$
$\Rightarrow$ radius $= 12 \cdot 5$ cm

$$\frac{\text{Area of}}{\text{plastic}} = \frac{\text{Outside}}{\text{circle area}} - \frac{\text{Inside}}{\text{circle area}}$$ ✓
$$= \pi \times 15^2 - \pi \times 12 \cdot 5^2$$ ✓
$$= \pi(15^2 - 12 \cdot 5^2)$$
$$= \pi(15 - 12 \cdot 5)(15 + 12 \cdot 5)$$
$$= \pi \times 2 \cdot 5 \times 27 \cdot 5$$
$$= 215 \cdot 98\ldots$$
$$\doteqdot 216 \text{ cm}^2$$ ✓
(to the nearest 1 cm²).

3 marks

Strategy
- Removal of the smaller circle area from the area of the larger outer circle is the crucial idea for the strategy mark.

Substitution
- The given diameter measurements should be halved to give the radius. This is necessary as the area formula $A = \pi r^2$ uses the radius value and not the diameter value.

Calculation
- The use of common factor and difference of squares is not expected — but it is fun!
- The alternative is to enter $\pi \times 15^2 - \pi \times 12 \cdot 5^2$ as written straight into your calculator.
SUCCESS GUIDE: page 38

Q6. $\sin x° \cos x° \tan x° + \cos^2 x°$

$= \sin x° \times \cos x° \times \dfrac{\sin x°}{\cos x°} + \cos^2 x°$ ✓

$= \sin^2 x° + \cos^2 x°$ ✓

$= 1$ ✓

3 marks

1st Identity
- Use $\tan x° = \dfrac{\sin x°}{\cos x°}$

Simplification
- Cancel the $\cos x°$ terms giving $\sin x° \times \sin x°$ which is written $\sin^2 x°$

2nd Identity
- Use $\sin^2 x° + \cos^2 x° = 1$
SUCCESS GUIDE: page 99

Q7. The equilateral triangle has three angles of $60°$ each:

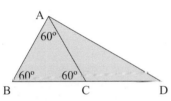

✓

Since the two pieces fit to form a triangle ABD then the edge BCD forms a straight angle of $180°$. ✓

So angle $ACD = 180° - 60° = 120°$.

Since triangle ACD is isosceles

$$\text{angle } CAD = \text{angle } CDA$$
$$= \tfrac{1}{2}(180° - 120°)$$
$$= 30°$$

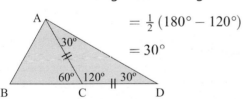

It is right-angled since:

angle $BAC +$ angle $CAD = 60° + 30° = 90°$

$\Rightarrow$ angle $BAD = 90°$ ✓

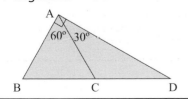

3 marks

Angles in an Equilateral Triangle
- No length are given so calculations involving angles are expected in your solution.

Strategy
- The crucial step in your solution is to realise that since the two smaller triangles fit together to form a larger triangle this means that the longer base edge is a straight line. If it was not a straight edge then the larger shape would not be a triangle.

Calculation and Conclusion
- You should make your reasoning clear throughout your solution or you may fail to earn all the marks available.
- Your conclusion should be clearly stated along with the reason:
  angle $BAC +$ angle CAD
  $= 60° + 30° = 90°$
SUCCESS GUIDE: pages 79–81

Q8. (a) $\dfrac{2}{3} - \dfrac{3x}{4} = \dfrac{x}{2}$

$\Rightarrow 12\left(\dfrac{2}{3} - \dfrac{3x}{4}\right) = 12 \times \dfrac{x}{2}$

$\Rightarrow \qquad 8 - 9x = 6x$ ✓

$\Rightarrow \qquad\qquad 8 = 15x$ ✓

$\Rightarrow \qquad\qquad x = \dfrac{8}{15}$ ✓

3 marks

Dealing with fractions
- Multiply both sides by 12 and cancel

Rearrangement
- Gather x terms on one side.
  Alternatively: $-15x = -8$

Solve for x
- Divide both sides by 15.
- Since the question specifically asks you for a fraction you will lose this mark for answers such as $0 \cdot 53$ or $0 \cdot 533$ etc
SUCCESS GUIDE: pages 31, 49

Q8. (b) Part (a) gives the x-coordinate of this point of intersection so substitute the value $x = \dfrac{8}{15}$ into either of the equations:

using the simpler equation $y = \dfrac{x}{2}$

using $y = \dfrac{\frac{8}{15}}{2} = \dfrac{\frac{8}{15} \times 15}{2 \times 15} = \dfrac{8}{30} = \dfrac{4}{15}$ ✓

1 mark

y-coordinate
- To find the point of intersection of the lines $y = \frac{2}{3} - \frac{3x}{4}$ and $y = \frac{x}{2}$ you solve $\frac{2}{3} - \frac{3x}{4} = \frac{x}{2}$ which you did in part (a) of the equation. So $x = \frac{8}{15}$ is the x-coordinate of this point.
- An alternative calculation is:
  $y = \frac{x}{2} = \frac{1}{2}x = \frac{1}{2} \times \frac{8}{15} = \frac{8}{30} = \frac{4}{15}$

SUCCESS GUIDE: page 16

Q9. (a)

Volume of cone $= \dfrac{1}{3}\pi r^2 h$

with $r = \dfrac{4}{2} = 2$ cm and $h = 6$ cm.

So volume $= \dfrac{1}{3} \times \pi \times 2^2 \times 6$ ✓

$= 25 \cdot 132 \ldots$ ✓

$\doteqdot 25 \cdot 1$ cm^3

(to 3 significant figures) ✓

3 marks

Substitution
- This 1st mark is for correctly substituting the values $r = 2$ and $h = 6$ into the cone formula.
- The formula is given to you on your formulae page during your exam.
- The radius (r) is used in the formula. You are given the diameter (4 cm) and so you need to halve this.

Calculation
- Remember to use the $\boxed{\pi}$ button (not $3 \cdot 14$).

Rounding
- There is a mark allocated for correct rounding to 3 significant figures. An answer of 25 cm^3 would not gain this mark — it's only got 2 significant figures!
SUCCESS GUIDE: page 41

Q9. (b)

Volume of a cylinder $= \pi r^2 h$

So volume of the weight

$$= \frac{1}{2}\pi r^2 h$$

where $r = 2$ cm.

Volume $= 25 \cdot 132\ldots$ cm³ and h is not known

$$\Rightarrow 25\cdot132\ldots = \frac{1}{2} \times \pi \times 2^2 \times h \quad \checkmark$$

$$\Rightarrow 25\cdot132\ldots = 2\,\pi \times h \quad \checkmark$$

$$\Rightarrow h = \frac{25\cdot132\ldots}{2\pi} = 4 \quad \checkmark$$

The height of the weight is 4 cm.

3 marks

Volume of weight

- Correct substitution in $\frac{1}{2}\pi r^2 h$ gains this mark

Equation

- Setting the volume of the weight equal to the volume of the cone calculated in part (a) is the strategy for finding h.
- Don't use 25·1 cm³ — this was rounded.

Solving

- Solve the equation to gain this mark
- You might wonder at exactly 4 cm being the solution. Is it exactly 4? Yes it is:

$$\frac{1}{3}\pi \times 2^2 \times 6 = \frac{1}{2}\pi \times 2^2 \times h$$

$$\left(\text{divide by } \pi \times 2^2\right)$$

$$\Rightarrow \frac{1}{3} \times 6 = \frac{1}{2}h \Rightarrow 2 = \frac{1}{2}h \Rightarrow h = 4$$

- This proof uses the 'exact' expression for the volume of the cone from part (a).

SUCCESS GUIDE: page 42

Q10. (a)

x	$x - \bar{x}$	$(x - \bar{x})^2$
503	2	4
504	3	9
497	−4	16
495	−6	36
506	5	25
		$\Sigma(x - \bar{x})^2 = 90$

$\checkmark$ (at row 497)

$$S = \sqrt{\frac{\Sigma(x - \bar{x})^2}{n - 1}} = \sqrt{\frac{90}{4}} \quad \checkmark$$

$$= 4\cdot743\ldots \doteqdot 4\cdot7 \quad \checkmark$$

(to 1 decimal place)

3 marks

Squared Deviations

- This mark is gained for correctly calculating the values in the last column of the table, i.e. $(x - \bar{x})^2$.
- No negative values should ever appear in this column. Remember when you square a quantity your answer will be positive or zero
- The mean, $\bar{x}$, has been given to you in the question and so you do not need to calculate it again!

Substitution

- This mark is for the correct substitution of the values $\Sigma(x - \bar{x})^2 = 90$ and $n - 1 = 4$ into the standard deviation formula.

Calculation

- Remember always that there is a square root to be taken at the end of a standard deviation calculation.
- Any reasonable rounding is acceptable.
- Using the other formula gives $\Sigma x = 2505$, $\Sigma x^2 = 1\,255\,095$ and $n = 5$ resulting in $s \doteqdot 4\cdot7$.

SUCCESS GUIDE: pages 130, 131

Q10. (b) Yes, it did. The new standard deviation of 3·5 is less than the previous value of 4·7 so there was less variation about the mean.   ✓

1 mark

Statement
- Your explanation must use the standard deviation statistics that you calculated in part (a) and that are also given in the question. The greater the standard deviation the greater the variation about the mean (and vice versa).

Q11. (a)

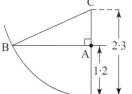

The radius of the fuselage is 2·3 m.
$AC = 2·3 - 1·2$
$= 1·1$ m (see diagram)   ✓

Use Pythagoras's theorem in triangle ABC:

$AB^2 = BC^2 - AC^2$ ✓
$= 2·3^2 - 1·1^2$
$= 4·08$
so $AB = \sqrt{4·08} = 2·019...$ ✓

By symmetry: width of floor
$= 2 \times AB$
so $x = 2 \times 2·019... = 4·039...$ ✓
$\Rightarrow x \doteq 4·04$ metres.

4 marks

Strategy
- Triangle ABC is right-angled and therefore Pythagoras's theorem can be used.

Pythagoras's theorem
- Notice that CB is a radius and that CA is part of a radius. Any radius in this circle has a length $\frac{1}{2} \times 4·6 = 2·3$ m

Calculation
- AB is one of the smaller sides in triangle ABC so the calculation involves a subtraction.
- Remember that rounding off decimal answers should only be done at the end of the problem when the final answer is given.

Solution
- By symmetry AB is half the width of the compartment floor.

Q11. (b) By symmetry (see diagram) the ceiling is as far above C as the floor is below C.

Height $= 2 \times 1·1$
$= 2·2$ metres   ✓

1 mark

Solution
- Two equal and parallel chords in a circle will be the same distance from the centre because the diagram is symmetric.

SUCCESS GUIDE: pages 76, 85

Q12. For the points of intersection
solve $\cos x° = 0.2$ ✓
$x°$ is in the 1st or 4th quadrants
1st quadrant angle is given by:
$x = \cos^{-1}(0.2) \div 78.5$ ✓
th quadrant angle is given by:
$x = 360 - 78.5 = 281.5$

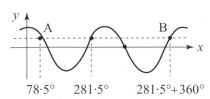

The x-coordinate of A is 78.5
From the graph above,
B lies 1 cycle on from the 4th
quadrant intersection.
The x-coordinate of B is $281.5 +$
$360 = 641.5$. ✓

3 marks

Strategy
- At each of the points of intersection the y-coordinate is 0·2 and the x-coordinate satisfies $\cos x° = 0.2$.
- The appearance of the equation $\cos x° = 0.2$ and an attempt to solve it will gain you this mark.

1st quadrant angle
- Using $\boxed{\cos^{-1}}$ on your calculator will always give you the 1st quadrant angle (don't use this key with a negative value).

Other angles
- You should identify the 4 quadrants on the graph that is given:

It should then be clear that the x-coordinate of B is given by the 4th quadrant value $+360°$
- The 4th quadrant value is obtained using this diagram:

$$180° - 1st \text{ quad}$$
$$180° + 1st \text{ quad} \mid 360° - 1st \text{ quad}$$

SUCCESS GUIDE: pages 96, 97

Q13. (a)
  4th line: $8 + 1 = 4 \times 6 - 5 \times 3$ ✓

1 mark

Pattern extension
- The order of writing the line cannot be altered.

Q13. (b)
1st line: $2 + 1 = 1 \times 3 - 2 \times 0$
2nd line: $4 + 1 = 2 \times 4 - 3 \times 1$
3rd line: $6 + 1 = 3 \times 5 - 4 \times 2$
4th line: $8 + 1 = 4 \times 6 - 5 \times 3$
                                                                ✓
nth line: $2n + 1 = n\,(n + 2) - (n + 1)\,(n - 1)$
                                                                ✓

2 marks

1st expression
- For finding one of $2n + 1$ or $n(n + 2)$ or $(n + 1)(n - 1)$ you will gain the 1st mark here.

Remaining expressions
- Careful use of brackets is essential:
  $n \times n + 2 (= n^2 + 2)$ is not the same as $n(n + 2)$
  $n + 1 \times n - 1 (= 2n - 1)$ is not the same as $(n + 1)(n - 1)$.
- Always check: for example substituting $n = 3$ should produce the 3rd line.

Q13. (c) Simplify the right-hand side of the nth line:

$n(n + 2) - (n + 1)(n - 1)$

$= n^2 + 2n - (n^2 - n + n - 1)$

$= n^2 + 2n - n^2 + n - n + 1$

$= 2n + 1$ ✓

This is the same as the left-hand side of the nth line.

So the pattern always holds.

1 mark

Simplification

- Again great care should be taken with brackets, especially expanding $(n + 1)(n - 1)$ because $-(n^2 - n + n - 1)$ is not the same as $-n^2 - n + n - 1$
- Proving $2n + 1 = n(n + 2) - (n + 1)(n - 1)$ using 'algebra' means that all lines are true. For instance, $n = 10$ gives: $20 + 1 = 10 \times 12 - 11 \times 9$ which is line 10.

SUCCESS GUIDE: pages 20, 21

Q14. (a) For 2 hours:

Earthmove: $64 + 2 \times 30 = £124$
  (delivery) (2 hours)

Trenchers: $28 + 2 \times 34 = £96$
  (delivery) (2 hours) ✓

1 mark

Calculation

- Both values being correct will gain you this mark.

Q14. (b) For 20 hours:

Earthmove: $64 + 20 \times 30$

$= £664$

Trenchers: $28 + 20 \times 34$

$= £708$ ✓

So Earthmove is cheaper in this case by £44

1 mark

Comparison

- The two calculations have to be clearly shown
- A comparison should be made between the two costs, e.g. '£44 cheaper' or '£664 is less than £708', etc.

Q14. (c) For n hours:

Earthmove: $64 + n \times 30$

$= £30n + 64$ ✓

Trenchers: $28 + n \times 34$

$= £34n + 28$ ✓

2 marks

1st formula

- Either one of the two expressions correctly stated will gain you the 1st mark here.

2nd formula

- Two correct expressions: two marks gained.

Q14. (d) For a 2 hour hire
Trenchers are cheaper, but not
for a 20-hour hire.
To find the 'cross-over value':
Solve $34n + 28 = 30n + 64$
$\Rightarrow 4n = 36 \Rightarrow n = 9$ ✓
Trenchers and Earthmove cost
the same for a nine-hour hire.
Any hire less than nine hours is
cheaper with Trenchers. ✓

2 marks

Strategy
- From parts (a) and (b) you should have realised that at some point Trenchers becomes the more expensive option having been cheaper when the number of hours was low.
- Using the formulae you worked out in (c) and setting them equal to each other would gain you this mark.
SUCCESS GUIDE: page 49

Solution and Interpretation
- What does $n = 9$ mean in the context of the question? A statement is necessary.

Q15. Triangles ABC and ADE are similar
(equiangular) ✓

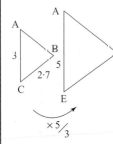

$AE = AC + CE$

$= 3 + 2$

$= 5 \text{ m}$

The enlargement

scale factor is $\dfrac{5}{3}$

so $DE = \dfrac{5}{3} \times 2 \cdot 7 = 4 \cdot 5 \text{m}$ ✓

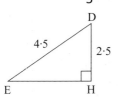

Use Pythagoras's
theorem in triangle
DEH: ✓
$EH^2 = ED^2 - DH^2$
$= 4 \cdot 5^2 - 2 \cdot 5^2$ ✓
$= 14$

so $GF = EH = \sqrt{14} = 3 \cdot 714...$
The gap between the vertical ✓
posts is $3 \cdot 71$ m

5 marks

Strategy
- Similar triangles calculation to find DE followed by Pythagoras's theorem in triangle DEH.

Similar Triangles
- Enlargements have scale factor greater than 1
- You should always 'disentangle' the two similar triangles, drawing them separately and writing in each of the lengths that you know
SUCCESS GUIDE: page 86

Pythagoras's theorem
- Finding a smaller side involves a subtraction.

Calculation
- Remember to use $\boxed{x^2}$ key in this calculation.

Answer
- Always check that your final answer seems reasonable in the context of the question. In this case $3 \cdot 71$ m compares well to just over 5 m for the 'king post' so it seems reasonable.
SUCCESS GUIDE: page 76